Global Climate Change and the Road to Extinction

Global Climate Change and the Road to Extinction

The Legal and Planning Response

James A. Kushner

Carolina Academic Press
Durham, North Carolina

Copyright © 2009
James A. Kushner
All Rights Reserved

Library of Congress Cataloging-in-Publication Data

Kushner, James A.
  Global climate change and the road to extinction / James A. Kushner.
      p. cm.
  Includes bibliographical references and index.
  ISBN 978-1-59460-492-8 (alk. paper)
  1. Climatic changes. 2. Global warming. I. Title.
  QC981.8.C5K875 2008
  363.738'74—dc22   2008035205

CAROLINA ACADEMIC PRESS
700 Kent Street
Durham, North Carolina 27701
Telephone (919) 489-7486
Fax (919) 493-5668
www.cap-press.com

Printed in the United States of America

[E]ven if every SUV were downsized to a Schwinn, every truck and bus repowered to burn biodiesel, and every refrigerator retrofitted to run with solar panels, we are playing Russian roulette with the very thing that makes our life on earth possible—a steady, temperate climate.

Jeff Goodell, Big Coal: The Dirty Secret Behind
America's Energy Future 207–08 (2006)

This book is dedicated to Al Gore, the world's spokesman and leader in addressing global climate change.

Contents

List of Figures	xiii
Preface	xv
Acknowledgments	xvii
Chapter 1 · Introduction	3
Chapter 2 · The Science and Consequences	7
Chapter 3 · Agriculture and Food Policy	19
1. Local Response	27
2. Regional Response	29
3. State Response	30
4. Federal Response	31
5. International Response	31
Chapter 4 · Brownfield Development	35
1. Local Response	37
2. Regional Response	37
3. State Response	38
4. Federal Response	38
5. International Response	39
Chapter 5 · Consumption and Conservation	41
1. Local Response	53
2. Regional Response	55
3. State Response	55
4. Federal Response	56
5. International Response	56
Chapter 6 · Economic Development	59
1. Local Response	62
2. Regional Response	63
3. State Response	64

4. Federal Response	64
5. International Response	66
Chapter 7 · Education	**69**
1. Local Response	71
2. Regional Response	72
3. State Response	72
4. Federal Response	72
5. International Response	73
Chapter 8 · Emergency Preparedness	**75**
1. Local Response	83
2. Regional Response	84
3. State Response	84
4. Federal Response	85
5. International Response	87
Chapter 9 · Energy	**89**
1. Local Response	102
2. Regional Response	103
3. State Response	103
4. Federal Response	111
5. International Response	112
Chapter 10 · Housing and Construction	**115**
1. Local Response	120
2. Regional Response	124
3. State Response	124
4. Federal Response	125
5. International Response	126
Chapter 11 · Management of Federal Lands and Agencies	**127**
1. Local Response	128
2. Regional Response	128
3. State Response	129
4. Federal Response	129
5. International Response	129
Chapter 12 · Oceans and Seas	**131**
1. Local Response	133
2. Regional Response	133
3. State Response	134

4. Federal Response	134
5. International Response	134
Chapter 13 · Population	**135**
1. Local Response	138
2. Regional Response	138
3. State Response	138
4. Federal Response	138
5. International	139
Chapter 14 · Smart Growth	**141**
1. Local Response	147
2. Regional Response	148
3. State Response	149
4. Federal Response	152
5. International Response	152
Chapter 15 · Species Protection	**153**
1. Local Response	155
2. Regional Response	155
3. State Response	156
4. Federal Response	156
5. International Response	156
Chapter 16 · Technology	**157**
1. Local Response	157
2. Regional Response	158
3. State Response	159
4. Federal Response	159
5. International Response	159
Chapter 17 · Transportation	**161**
1. Local Response	182
2. Regional Response	186
3. State Response	186
4. Federal Response	189
5. International Response	191
Chapter 18 · Water Management	**195**
1. Local Response	198
2. Regional Response	199
3. State Response	199

4. Federal Response	200
5. International Response	201
Chapter 19 · Conclusion	203
Chapter 20 · Afterword	207
Table of Authorities	211
Table of Statutes, Constitutional Provisions, Regulations, and Executive Orders	243
Table of Cases	251
Index	253

List of Figures

Polar bears. Source: istockphoto.com/Thomas Pickard.	front cover
Flooding from Hurricane Katrina, New Orleans, Sept. 4, 2005. Source: FEMA/Liz Roll.	5
U.S. Carbon Dioxide Emissions 2005 to 2030. Source: Department of Energy.	10
Tornado, Union City, Oklahoma. Source: National Oceanic & Atmospheric Administration.	13
Cattle Feed Yard. Source: USDA.	21
Poultry Production. Source: USDA/Larry Rana.	26
Millennium Housing, dockyards to wetlands and housing, London. Source: James A. Kushner.	36
Energy Star compact fluorescent light bulb (CFL). Source: EPA/DOE.	54
Bluewater Mall in Kent, UK. Source: James A. Kushner.	57
Tram in Budapest, Hungary. Source: James A. Kushner.	62
Student. Source: U.S. Census Bureau, Public Information Office.	71
Superdome line, New Orleans, Aug. 28, 2005. Source: FEMA/Marty Bahamonde.	77
Evacuation from Galveston for Hurricane Rita, Sept. 21, 2005. Source: FEMA/Ed Edahl.	79
Thames Barrier, London. Source: James A. Kushner.	82
Geysers Geothermal Power Plant, California. Source: DOE.	92
Heliostats at the Solar Two Power Plant, Daggett, California. Source: DOE.	98
Flowind, Altamont Pass, California. Source: DOE.	99
Bio-01 sustainable community (West Harbor), Mälmo, Sweden. Source: James A. Kushner.	117
Autofreie Siedlung carfree housing, Munster, Germany. Source: James A. Kushner.	119
WGL-Terrein car-free housing, Amsterdam, NL. Source: James A. Kushner.	121

Zero Energy Houses, The Vauban, Freiburg, Germany. Source:
James A. Kushner. 123
Wind Generators on Federal Lands in Palm Springs, California.
Source: U.S. Bureau of Land Management. 128
Huge surf with offshore wind North Shore, Oahu, Hawaii.
Source: National Oceanic & Atmospheric Administration. 133
World population projection. Source: U.S. Census Bureau. 136
French Quarter, New Urbanist village, Tübingen, Germany.
Source: James A. Kushner. 146
Polar bear. Source: NOAA Climate Program Office NABOS 2006
Expedition/Mike Dunn. 154
University of Minnesota Solar car passes through Lake Benton,
Minnesota. Source: U.S. Department of Energy. 158
Traffic, Los Angeles. Source: Richard Riesemberg, www.rickrise.com. 162
Hydrogen-fueled bus, Estoril, Portugal. Source: Carfree.com. 163
Tram in Bratislava, Slovakia. Source: James A. Kushner. 166
Double articulated bus on dedicated busway at Utrecht University in
The Netherlands. Source: James A. Kushner. 170
Bicycles, Amsterdam, NL. Source: James A. Kushner. 171
Smart Car, Wurtzburg, Germany. Source: James A. Kushner. 173
Bike Rental, Vienna, Austria. Source: James A. Kushner. 175
Richshaw bicycle, Copenhagen, Denmark. Source: James A. Kushner. 177
Articulated bus, Nuremberg, Germany. Source: James A. Kushner. 181
Tram, Freiburg, Germany. Source: James A. Kushner. 184
New Orleans, Sept. 6, 2005. Source: FEMA/Jocelyn Augustino. 204
Marble Road, Ephesus, Turkey. Source: James A. Kushner. 205
New Orleans, Sept. 6, 2005. Source: FEMA/Jocelyn Augustino. back cover

Preface

In the spring of 2007, while living and teaching again in the Netherlands, I went to see "An Inconvenient Truth," Nobel Peace Prize recipient Al Gore's Academy Award winning film on global warming and climate change. While watching the film, I had an epiphany: global warming is a threat to life as we know it, eclipsing all other global, national, and local concerns and that urban planning and urban redesign can significantly mitigate the effects of climate change. I had been aware of global warming for some years, particularly as to the danger to low-lying lands from rising seas. But as I began to read everything I could find on warming and greenhouse gases, starting with Al Gore's book *An Inconvenient Truth*, I realized that Earth is on a destructive path and with it, mankind is heading for early extinction. In this book, I do critique Al Gore's support for the Kyoto Protocol and his modest prescriptive and normative response to climate change. In fairness, we are moving quickly toward a crisis and our knowledge of the effects of warming is expanding rapidly. I am sure that Al Gore also recognizes that more recent information calls for even more aggressive policies to mitigate the effects of our ever-increasing emissions of greenhouse gases. I applaud his leadership and am proud to acknowledge that I have joined his army of fighters for sustainability and survival.

I am not a scientist, but even the most moderate experts are predicting a scenario that has already adversely affected species, changed disease patterns, and increased flooding and, in the coming decade, will cause water and food shortages, hundreds of millions of deaths, and climate refugees to be followed by climate conditions and economic chaos that will threaten the quality of life and the very life of our children and grandchildren. I do not believe that political leaders in the United States and abroad want to face the threat of climate change because they fear that such honesty would shake economies and political futures. Instead, the response is a mild commitment to increase miles per gallon automobile efficiency and efforts to stimulate alternatives to fossil fuel-based economies. Many people assume that scenarios of devastation are science fiction and thus respond by ignoring scientific forecasts. The premise of this volume is that the risk forecasted is so grave that it is only prudent to look to the future and make the changes in urban design and consumption necessary to avert extinction and move to a more sustainable existence. Few dispute the existence of peak oil and the coming crisis of insufficient supply and escalating costs of oil, natural gas, coal, and nuclear energy and that we should plan to convert to renewable energy. No one should dispute that our current policies are subsidizing pollution and carbon emissions. No one should dispute that we could generate an economy based on conservation, creating a new generation of jobs and opportunities. The point of this book is not only to avert extinction should dire predictions be correct, but to take the crisis of global climate change and treat it as an opportunity to launch the sustainable lifestyle and economy as a gift to our progeny. It would be ironic and tragic if this generation, a generation which has been more protective of its children than any generation before it, would also ignore the greatest threat to them.

Chapter 1

Introduction

> There may be some excuses for great planning disasters, but there are not nearly as many as we think.
>
> PETER HALL, GREAT PLANNING DISASTERS 276 (1980)

> Modern Technology—primarily the automobile—and contemporary economic, cultural, and social forces continue to erode and spread out the city. Not only is the modern metropolis at risk, modernism itself is in deep trouble.
>
> DOUGLAS KELBAUGH, COMMON PLACE: TOWARD NEIGHBORHOOD AND REGIONAL DESIGN 3 (1997)

> The urgency of the climate problem, the ever-increasing scientific certainty that "business as usual" will lead to irreversible, unacceptable outcomes, undermines the deep-seated analytical presumption in favor of the status quo. What climate science tells us, above all, is that the status quo is not going to remain one of the available options.
>
> Lisa Heinzerling & Frank Ackerman, *Law and Economics for a Warming World*, 1 HARV. L. & POL'Y REV. 331, 333 (2007)

Global warming and global climate change are already a reality threatening the lives of hundreds of millions, extinction of uncountable species of plants, animals, and fish, and very possibly the extinction of man.[1] The Nobel prize winning International Panel on Climate Change (IPCC) in its final report of 2007 stated that climate change is now a "certainty" and warned that unless

---

1. 3 BRUCE E. JOHANSEN, GLOBAL WARMING IN THE 21ST CENTURY: PLANTS AND ANIMALS IN PERIL (2006); Frederic H. Wagner, *Global Warming Effects on Climactically-Imposed Ecological Gradients in the West*, 27 J. LAND RESOURCES & ENVTL. L. 109 (2007); Associated Press, *Climate Report Sound Dire Warnings: Global Warming Could Mean Hundreds of Millions without Water*, MSNBC.COM, Mar. 10, 2007 (based on draft of climate report). *See* also Chap. 2.

action is taken, human activity could lead to "abrupt and irreversible changes" that would make the planet unrecognizable.[2] Climate is not just about temperature, but includes precipitation, wind, and other climate characteristics; however, warming is objective and quantifiable, while climate is less so.[3] I use the term climate change to include temperature change. As nations and corporations discuss and undertake reform measures to slow or reverse this warming trend, the questions arise: What strategies exist to reverse the problems caused by greenhouse gases collecting in our atmosphere? What steps can be taken to reduce emissions? This book looks at strategies that can, and in most cases must, be undertaken at the personal, corporate, and governmental levels of municipalities, counties, states, nations, and within the international community.

Al Gore, in his book "An Inconvenient Truth," sets out six changes to reduce emissions to a point below 1970 levels.[4] The changes include (1) using electricity more efficiently; (2) designing buildings and businesses to use less energy; (3) increasing automobile and truck efficiency by using less gas and more hybrids and alternative fuels; (4) designing cities and towns around mass transit; (5) increasing reliance on renewable energy technologies; and (6) capturing and storing excess carbon from power plants and industrial activities.

This book is designed to look at broad changes that are necessary and to explore the specific policies and mechanisms that are needed to implement them. The discussion and recommended policies offered will also critique the Gore proposed changes as they relate to energy consumption, public transport, motor vehicles, and construction. I have elected to focus on feasible options based on existing technology and have eschewed strategies based on exotic untested fuels from other planets, proposals of economists suggesting that par-

---

2. Msnbc, *U.N. Issues Landmark Report on Global Warming: Panel Offers, Establishes Scientific Baseline for Political Talks*, MSNBC, Nov. 17, 2007, available at http://www.msnbc.msn.com/id/21844627/ (Synthesis Report). *See also* Elisabeth Rosenthal, *U.N. Report Describes Risks of Inaction on Climate Change*, N.Y. TIMES, Nov. 17, 2007, available at http://www.nytimes.com/2007/11/17/science/earth/17climate.html?_r=1&ref=todayspaper&oref=slogin.

3. ANDREW E. DASSLER & EDWARD A. PARSON, THE SCIENCE AND POLITICS OF GLOBAL CLIMATE CHANGE: A GUIDE TO THE DEBATE 47–48 (2006).

4. AL GORE, AN INCONVENIENT TRUTH: A PLANETARY EMERGENCY OF GLOBAL WARMING AND WHAT WE CAN DO ABOUT IT 281 (2006). *See also* TIM FLANNERY, THE WEATHERMAKERS: HOW MAN IS CHANGING THE CLIMATE AND WHAT IT MEANS FOR LIFE ON EARTH 303–06 (2005); *Top 10 Things You Can Do to Reduce Global Warming*, available at http://environment.about.com/od/globalwarming/tp/globalwarmtips.htm.

Flooding from Hurricane Katrina, New Orleans, Sept. 4, 2005.
Source: FEMA/Liz Roll.

ticles be added to the atmosphere to scatter a small portion of incoming sunlight back into space,[5] or futuristic plans to transmit solar power to earth from gigantic solar panels in space.[6]

Every facet of life must be reviewed for sustainability to determine opportunities for reducing carbon emissions through alternative technology and conservation practices. The following chapters will explore these opportunities, and discuss the significance of our policies on agricultural and food, land use planning, brownfield development, redevelopment, consumption, economic development, education, emergency preparedness, energy, housing and construction, management of federal lands, seas, population, species protection, transportation, and water.

Although some argue it is already too late to act to save our civilization, a myriad of already available and soon to be available strategies and mechanisms

---

5. Alan Carlin, *Global Climate Change Control: Is There a Better Strategy Than Reducing Greenhouse Gas Emissions?*, 155 U. PA. L. REV. 1401 (2007).

6. Charles J. Hanley, *'Drilling Up' Into Space for Energy*, ASSOCIATED PRESS, Dec. 23, 2007, *available at* http://ap.google.com/article/ALeqM5gMOg-D8-UyHFE3GmhgU5eUM-RVF0gD8TNBC2G0.

could be implemented that would make a significant difference and may even save the planet as we know it.

Chapter 2

The Science and Consequences

> The history of life on earth has been a history of interaction between living things and their surroundings. To a large extent, the physical form and the habits of the earth's vegetation and its animal life have been molded by the environment. Considering the whole span of earthly time, the opposite effect, in which life actually modifies its surroundings, has been relatively slight. Only within the moment of time represented by the present century has one species—man—acquired significant power to alter the nature of his world.
>
> Rachel Carson, Silent Spring 5 (1962)

Although the media may have suggested that the existence of global warming is controversial and contested,[7] reflecting an organized campaign of corporate

---

7. Mark Jaffe, *Global Warming?*, Denver Post, Dec. 26, 2006 (disputed); *Cheney on Global Warming: Vice President's Views at Odds with Majority of Climate Scientists*, ABC News Technology and Science, Feb. 23, 2007) (acknowledging existence but suggesting it may be natural cycle rather than caused by humans); *Global Warming*, Business Week On Line, Aug. 16, 2004 (growing consensus). *See also* Gore, *supra* note 4 at 260–63 (reporting that while 928 scientific reports during the last 10 years in agreement on the cause of global warming and none to the contrary, 53% of popular press stories during the same period expressed doubt on the cause); Andrew C. Revkin, *Climate Change as News: Challenges in Communicating Environmental Science*, in Climate Change: What it Means for Us, Our Children, and Our Grandchildren 139, 151–54 (Joseph F.C. DiMento & Pamela Doughman eds. 2007); Richard Wolfson and Stephen H. Schneider, *Understanding Climate Science, in* Climate Change Policy: A Survey 3, 41–42 (Stephen H. Schneider, Armin Rosencranz & John O. Niles eds. 2002).

disinformation,[8] the science is quite simple and uncontested[9] except for a few popular fiction writers,[10] a few skeptic scientists,[11] and lots of people in the United States who do not know or do not want to know. Perhaps not coincidentally, a number of the lonely doubters are meteorologists who have the unhappy challenge of rewriting forecast models since much of their work and

---

8. JEFF GOODELL, BIG COAL: THE DIRTY SECRET BEHIND AMERICA'S ENERGY FUTURE 180–81 (2006) (describing how American public opinion on fears of global warming dropped by one-third from 35% to 22% as a result of a campaign by the Western Fuels Association, a group of coal-burning utilities involving lectures, film, and press contacts); Sharon Begley, *The Truth About Denial,*. NEWSWEEK, Aug. 13, 2007, *available at* http://www.msnbc.msn.com/id/20122975/site/newsweek/page/0/.

9. INTERGOVERNMENTAL PANEL ON CLIMATE CHANGE, CLIMATE CHANGE 2001 93, *available at* http://grida.no/climate/ipcc_tar/wg2/index.htm; Jeffrey Kluger, *Global Warming Heats Up*, TIME, Mar. 26, 2006; Naomi Oreskes, *Beyond the Ivory Tower: The Scientific Consensus on Climate Change*, 306 SCIENCE 1686, 1686 (2004) (surveying 928 scientific, peer-reviewed journal articles, and finding that none disagreed with the consensus position that human activities are responsible for dramatic increases in greenhouse gases in the atmosphere, which in turn have caused surface air temperatures and subsurface ocean temperatures to rise.); Naomi Oreskes, *The Scientific Consensus on Climate Change: How Do We Know We're Not Wrong?*, in CLIMATE CHANGE: WHAT IT MEANS FOR US, *supra* note 7 at 65. *See also* Massachusetts v. EPA, 127 S. Ct. 1438 (2007) (Clean Air Act authorizes the EPA to regulate greenhouse gas emissions from new motor vehicles in the event that it forms a "judgment" that such emissions contribute to climate change which is accepted by most experts); Lisa Heinzerling, *Climate Change and the Clean Air Act*, 42 U.S.F.L. REV. 111 (2007) (arguing for EPA authority to regulate carbon and other greenhouse gases sustained in *Massachusetts v. EPA*).

10. MICHAEL CRICHTON, STATE OF FEAR (2004). *See also* ANN COULTER, GODLESS 188–92 (2006); 1 BRUCE E. JOHANSEN, GLOBAL WARMING IN THE 21ST CENTURY: OUR EVOLVING CLIMATE CRISIS 124–45 (2006); JOSEPH J. ROMM, HELL AND HIGH WATER: GLOBAL WARMING — THE SOLUTION AND THE POLITICS — AND WHAT WE SHOULD DO 212–229 (2007). George F. Will, *Fuzzy Climate Math*, WASH. POST, Apr. 12, 2007, at A27; http://www.skepticism.net/faq/environment/global_warming/index.html (writings skeptical of global warming).

11. GOODELL, BIG COAL, *supra* note 8 at 195–97 (describing the 1998 Oregon Petition purporting to have been signed by 17,000 scientists disputing human-impacted global warming which a sample by *Scientific American* found likely to contain perhaps 200 climate researchers, many of which claim to have never signed or would not sign today); MARCEL LEROUX, GLOBAL WARMING — MYTH OR REALITY? THE ERRING WAYS OF CLIMATOLOGY (2005); Christina K. Harper, *Climate Change and Tax Policy*, 30 B.C. INT'L & COMP. L. REV. 411, 415–16 (2007) (describing skeptics); S. FRED SINGER & DENNIS T. AVERY, UNSTOPPABLE GLOBAL WARMING: EVERY 1,500 YEARS (2007). For critiques of the skeptics, see DASSLER & PARSON, *supra* note 3 at 135–54 *and* ROSS GELBSPAN, THE HEAT IS ON: THE HIGH STAKES BATTLE OVER EARTH'S THREATENED CLIMATE 197–237 (1997) (critique of skeptics) *and* GEORGE MONBIOT, HEAT: HOW TO STOP THE PLANET FROM BURNING 20–42 (2007) ("the denial industry").

science would be invalid given the rapidly changing world.[12] Greenhouse gases such as carbon dioxide, methane, nitrous oxide, and chloroflurocarbons (CFCs) are increasing in the earth's atmosphere.[13] Carbon is of primary concern because while other greenhouse gases are often destroyed by chemical reaction in the atmosphere, carbon dioxide is not.[14] Atmospheric carbon has increased by 300% between 1900 and 2000.[15] This dramatic increase in carbon dioxide, compared to the increase in other greenhouse gases, is the reason it is the target in the fight to stabilize and reduce global warming.[16]

Carbon dioxide levels are rising 200 times faster now than they have at any time in the last 650,000 years.[17] Each year, 5.4 billion tons are added to the atmosphere from burning fossil fuels and another 1.6 billion tons are added from deforestation and the resultant loss of sequestration by trees.[18] Deforestation contributes as much as 20 to 25% of annual carbon emissions.[19] Hurricane Katrina resulted in the destruction of 320 million trees, which will ultimately release 367 million tons of carbon dioxide as they decompose.[20] This release is equal to all of the carbon absorbed by forests in the U.S. annually and exceeds the entire season's worth of emissions from U.S. forest fires.[21] Compared to the 1980 eruption of Mt. St. Helens that destroyed 150,000 acres of forest, Katrina damaged 5 million acres of forests.[22] In 2002, the burning of fossil fuels released 23 billion tons of carbon dioxide into the atmosphere.[23] Presently, the

---

12. Myles Allen *et al.*, *Scientific Challenges in the Attribution of Harm to Human Influence on Climate*, 155 U. PA. L. REV. 1353, 1356 (2007).

13. JOHN HOUGHTON, GLOBAL WARMING: THE COMPLETE BRIEFING 22–43 (2d ed. 1997). *See also* GORE, *supra* note 4 at 28.

14. HOUGHTON, *supra* note 13 at 24.

15. Frederick A.B. Meyerson, *Population and Climate Change Policy*, *in* CLIMATE CHANGE POLICY: A SURVEY 251, 251 (Stephen H. Schneider, Armin Rosencranz & John O. Niles eds. 2002).

16. HOUGHTON, *supra* note 13 at 1–9. *See also* GORE, *supra* note 4 at 30–37, 66–67 (nearly 400 parts per million compared to a high of 260 during the last 100,000 years).

17. ROMM, HELL AND HIGH WATER, *supra* note 10 at 20.

18. David S. Chapman & Michael G. Davis, *Global Warming—More than Hot Air?*, 27 J. LAND RESOURCES & ENVTL. L. 59, 67 (2007). *See generally* David J. Hayes & Joel C. Beauvais, *Carbon Sequestration*, in GLOBAL CLIMATE CHANGE AND U.S. LAW 691 (Michael B. Gerrard ed. 2007).

19. UNITED NATIONS ENVIRONMENT PROGRAMME, GEO4—GLOBAL ENVIRONMENT OUTLOOK 49 (2007), *available at* http://www.unep.org/geo/geo4/report/GEO-4_Report_Full_en.pdf.

20. Thomas H. Maugh II & Karen Kaplan, *Katrina Leaves Permanent Scar on Forests*, L.A. TIMES, Nov. 16, 2007, *available at* http://www.latimes.com/news/science/la-sci-trees16nov16,1,2189243.story.

21. *Id.*

22. *Id.*

23. FLANNERY, *supra* note 4 at 70.

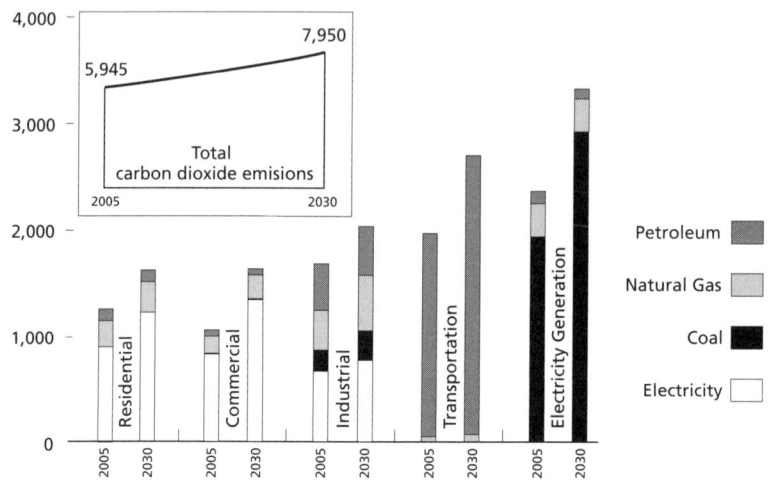

U.S. Carbon Dioxide Emissions 2005 to 2030 (millions of metric tons).
Source: Department of Energy.

burning of coal generates 41% of the total carbon dioxide emissions, 31% comes from oil, and 20% comes from gas.[24] In the United States, between 1960 and 2000, carbon dioxide emissions increased by 2.1% annually.[25] Four tons of carbon dioxide are released into the atmosphere for each ton of coal burned.[26]

The increase of greenhouse gases prevents the sun's radiation from escaping the atmosphere, thus increasing the earth's surface temperature.[27] Although greenhouse gas accumulation might suggest protection from the sun's rays, the effect of the extraordinary carbon gas accumulation on the planet Venus has raised its surface temperature from 50° Celsius (C) to 525° C.[28]

The increasing temperature is heating oceans, threatening sea life, melting glaciers,[29] reducing potable water,[30] and generating larger and more erratic cli-

---

24. *Id.*
25. MATTHEW E. KAHN, GREEN CITIES: URBAN GROWTH AND THE ENVIRONMENT 135 (2006).
26. FLANNERY, *supra* note 4 at 70.
27. HOUGHTON, *supra* note 13 at 10–20, 22–43.
28. *Id* at 16–20.
29. GORE, *supra* note 4 at 42–65; ROBERT HENSON, THE ROUGH GUIDE TO CLIMATE CHANGE: THE SYMPTOMS, THE SCIENCE, THE SOLUTIONS 71–99 (2006); 2 BRUCE E. JOHANSEN, GLOBAL WARMING IN THE 21ST CENTURY: MELTING ICE AND WARMING SEAS 443–62 (2006).
30. Associated Press, *Climate Report Sound Dire Warnings: Global Warming Could Mean Hundreds of Millions without Water*, MSNBC.COM, Mar. 10, 2007 (based on draft of cli-

mate change patterns.³¹ As the permafrost melts, methane is emitted increasing warming and melting.³² Warming will also accelerate the hydrologic cycle causing more evaporation and more atmospheric moisture but with some areas drier and some with more precipitation.³³ In addition, the increased precipitation and changes in the location of precipitation is generating desertification as well as flooding, and is a significant cause of changes in soil moisture and famine.³⁴ The decline in absorption of carbon dioxide by oceans aggravates the problem of rising temperature. Oceans absorbed 2 gigatons of carbon per year during the 1980s, but that sequestration effect declined to 1.8 gigatons in the 1990s.³⁵ Melting icecaps threaten to continue the rise in sea level and to generate flooding.³⁶ The incidence, intensity, and duration of storms has increased dramatically in recent years.³⁷ Global warming results in redistributing bacteria, insects, and diseases to higher altitudes and previously colder climates.³⁸ Warming also accelerates the breeding and biting of insects. Consider that malaria currently kills 1 million people and is causing 300 million acute illnesses each year.³⁹ In Africa it kills 3,000 children daily, a rate of

---

mate report predicting hundreds of millions of Africans and tens of millions in Latin America will be short of water).

31. 2 JOHANSEN, *supra* note 29 at 457–62 (El Niño).

32. JAMES MARTIN, THE MEANING OF THE 21ST CENTURY 17 (2006).

33. Kathleen A. Miller, *Climate Change and Water in the West: Complexities, Uncertainties and Strategies for Adaptation*, 27 J. LAND RESOURCES & ENVTL. L. 87 ns 16–17 and accompanying text (2007).

34. GORE, *supra* note 4 at 106–21. See also HOUGHTON, *supra* note 13 at 122, 125 (currently 60,000 square kilometers or 10,000 square miles are converted to desert annually; agricultural production likely to rise in developed northern hemisphere and decline in the south).

35. FLANNERY, *supra* note 4 at 35.

36. LESTER R. BROWN, PLAN B 2.0: RESCUING A PLANET UNDER STRESS AND A CIVILIZATION IN TROUBLE 66–74 (2006); DASSLER & PARSON, *supra* note 3 at 90–102. HOUGHTON, *supra* note 13 at 50; 2 JOHANSEN, *supra* note 1 at 405–441; ROMM, HELL AND HIGH WATER, *supra* note 10 at 87–95 (20- to 40-inch rise by 2050, 60- to 80-inch rise by 2100). See also A. BARRIE PITTOCK, CLIMATE CHANGE: TURNING UP THE HEAT 95–99 (2005) (sea-level rise).

37. BROWN, *supra* note 36 at 74–77; GORE, *supra* note 4 at 80–105; ROMM, HELL AND HIGH WATER, *supra* note 10 at 37–52; Robert L. Glicksman, *Global Climate Change and the Risks to Coastal Areas from Hurricanes and Rising Sea Levels: The Costs of Doing Nothing*, 52 LOY. L. REV. 1127 (2006) (arguing a growing consensus that hurricanes and storms are worsening from global warming and that the federal government is failing to address needed policy change).

38. Jia-Rui Chong, *Global Warming: Enough to Make You Sick*, LOS ANGELES TIMES, Feb. 25, 2007; 3 JOHANSEN, *supra* note 1 at 601–20.

39. ROSS GELBSPAN, BOILING POINT: HOW POLITICIANS, BIG OIL AND COAL, JOURNAL-

one child every 30 seconds.[40] Mosquitoes have now spread to previously colder climates and higher altitudes.[41]

Changing climate may modify the hatching times of birds and other species as well as the availability of food and the new schedules may deprive certain species of sustenance and possible survival.[42] Forests may be lost as fewer frosts may fail to eradicate predator insects such as the pine or bark beetle.[43] Warming trends point toward the increase in wildfires.[44] Global warming is exacerbated by the continued pattern of deforestation.[45] In addition, with an increase in temperature of 4° F expected by the end of the century,[46] forests will begin to die and cease absorbing carbon dioxide and so contribute even more to increasing temperature.[47] The increase in disease, heat waves, and the decline in forests, agricultural production and water poses a threat of suffering, refugees, and international strife.[48]

Experts estimate that global warming may cost the United States between $61 billion and $74 billion annually.[49] Warming impacts wind patterns as well. For

---

ISTS, AND ACTIVISTS ARE FUELING THE CLIMATE CRISIS—AND WHAT WE CAN DO TO AVERT DISASTER 122 (2004).

40. *Id.*

41. *Id* at 119–25. *See also New Virus Extends Geographic Reach*, L.A. TIMES, Dec. 8, 2007, at A13 (reporting Chikungunya virus spreading to Europe and North America by mosquitoes due to global warming).

42. GORE, *supra* note 4 at 152–53 (describing how the caterpillar season now precedes bird-hatching season when the source of food for chicks is no longer available).

43. *Id* at 154–57.

44. Eleanor G. Turman, *Regional Impact Assessments: A Case Study of California*, in CLIMATE CHANGE POLICY: A SURVEY 89, 98 (Stephen H. Schneider, Armin Rosencranz & John O. Niles eds. 2002).

45. John O. Niles, *Tropical Forests and Climate Change*, in CLIMATE CHANGE POLICY: A SURVEY ch. 13 (Stephen H. Schneider, Armin Rosencranz & John O. Niles eds. 2002).

46. Andrew C. Revkin, *Budgets Falling in Race to Fight Global Warming*, N.Y. TIMES, Oct. 30, 2006, at A1 (predicting a rise of 4° Fahrenheit).

47. MARTIN, THE MEANING OF THE 21ST CENTURY, *supra* note 32 at 17.

48. Meyerson, in CLIMATE CHANGE POLICY, *supra* note 15 at 252. *See also* Kenneth D. Frederick & Peter H. Gleick, *Potential Impacts on U.S. Water Resources*, in CLIMATE CHANGE: SCIENCE, STRATEGIES, & SOLUTIONS 63–78 (Eileen Claussen, Vicki Arroyo Cochran & Debra P. Davis eds. 2001. *See also* Ruth Gordon, *Climate Change and the Poorest Nations: Further Reflections on Global Inequality*, 78 U. COLO. L. REV. 1559 (2007) (recognizing that the poor nations will suffer the worst from warming and arguing for a clean development fund mechanism to provide aid for those nations); Rebecca Tsosie, *Indigenous People and Environmental Justice: The Impact of Climate Change*, 78 U. COLO. L. REV. 1625 (2007) (arguing that policies of adaptation to climate change are genocidal to indigenous people and advocating a right to environmental determinism).

49. HOUGHTON, *supra* note 13 at 134.

Tornado, Union City, Oklahoma.
Source: National Oceanic & Atmospheric Administration.

each 10 knot rise of wind speed during a storm, building damage increases by 650%.[50] As of 2005, insurance losses from weather-related disasters has reached $200 billion annually.[51] By 2050, the worldwide cost from climate change is anticipated to reach $500 billion and by 2065, the annual damage cost will be equal to the value of everything that humanity produces in the course of a year.[52] Insurers are increasingly withdrawing from writing policies near the hurricane-prone Atlantic coast or imposing a 2–5% deductible on wind claims

---

50. *See also* FLANNERY, *supra* note 4 at 236.

51. 3 JOHANSEN, *supra* note 1 at xix–xx (comparing losses of $56.6 billion from hurricanes Katrina, Rita, and Wilma and $35 billion for losses from the destruction of the World Trade Center in 2001). *See also* Patrick E. Tolan, Jr., *Tax and Insurance Consequences of Major Disasters: Weathering the Storm*, 31 NOVA L. REV. 487 (2007) (describing tax subsidies and relief for victims and the need for insurance coverage); Sara Elizabeth Graditor, Comment, *Responsibility for the Restoration of the Hurricane Insurance Industry: Business Proposal or State Solution?* 31 NOVA L. REV. 527 (2007) (advocating state establishing hurricane and other loss insurance).

52. *See also* FLANNERY, *supra* note 4 at 235–36.

within two miles of the coast.⁵³ Although insurers are beginning to factor climate change and its consequences into their future policies, governments are failing to take global climate change seriously. As a consequence of this failure, economists are dismissing predictions and discounting the possibility of future disaster.⁵⁴ The existing pollution and its related health costs are costing the planet $300 billion annually.⁵⁵ The largest insurance company in the world, Munich Reinsurance, prices the direct losses from global climate change at $300 billion in a few decades.⁵⁶ Britain's largest insurance company predicts that unchecked climate change could bankrupt the global economy by 2065.⁵⁷ The cost to the United States is $20 billion annually but is projected to increase to $100 billion by 2050.⁵⁸ By 2050, the cost of global warming to the United States could reach $300 billion annually.⁵⁹ In comparison, the cost of reducing carbon dioxide is between $7 and $14 per ton, a cost that is achievable through existing energy and conservation technology.⁶⁰ The savings from slowing global warming to business and consumer energy bills would exceed the economic cost of doing so.⁶¹

Global warming and climate change has been succinctly and dramatically described and explained.⁶² Although the purpose of this work is not to cri-

---

53. Jennifer Couzin, *Living in the Danger Zone*, 319 SCIENCE 748, 748 (Feb. 2008); Bruce Mohl, *More Insurers Backing Away From Coasts*, BOSTON GLOBE, Oct. 14, 2007, available at http://www.boston.com/business/personalfinance/articles/2007/10/14/more_insurers_bac. *See generally* Michael G. Faure, *Insurability of Damage Caused by Climate Change: A Commentary*, 155 U. PA. L. REV. 1875 (2007); Howard C. Kunreuther & Erwann O. Michel-Kerjan, *Climate Change, Insurability of Large-Scale Disasters, and the Emerging Liability Challenge*, 155 U. PA. L. REV. 1795 (2007); Sarah M. Tran, *Updated Hurricane Models: A New Opportunity to Insure Against Climate Change*, 14 B.U. J. SCI. & TECH. L. 73 (2008) (arguing that insurers are using outdated hurricane predication models and grossly underestimating losses and should reduce insurability leaving those on the coasts to internalize risk).

54. Lisa Heinzerling & Frank Ackerman, *Law and Economics for a Warming World*, 1 HARV. L. & POL'Y REV. 331, 349–53 (2007).

55. Orie L. Loucks, *Business Capitalizing on Energy Transition Opportunities*, *in* CLIMATE CHANGE POLICY: A SURVEY 495, 496–97 (Stephen H. Schneider, Armin Rosencranz & John O. Niles eds. 2002).

56. MARTIN, THE MEANING OF THE 21ST CENTURY, *supra* note 32 at 104.

57. *Id*, citing GELBSPAN, BOILING POINT, *supra* note 39 at 112.

58. Loucks, *in* CLIMATE CHANGE POLICY, *supra* note 55 at 496–97.

59. Amy Cortese, *As the Earth Warms, Will Companies Pay?*, N.Y. TIMES, Aug. 18, 2002, §3, at 6.

60. JOSEPH J. ROMM, THE HYPE ABOUT HYDROGEN: FACT AND FICTION IN THE RACE TO SAVE THE CLIMATE 144 (2005).

61. *Id.*

62. GORE, *supra* note 4. *See generally* GARY BRAASCH, EARTH UNDER FIRE: HOW GLOBAL WARMING IS CHANGING THE WORLD 7–8 (2007); PAUL BROWN, GLOBAL WARNING: THE LAST

tique the science, I would argue that even if uncertainty as to the future exists, the risk is certain and catastrophic.[63] Two thousand scientists and 120 governments have announced that the planet is on the road to extinction.[64] Forecasts for temperature change are conflicting but nevertheless disturbing. Warming of the Earth by 1° C is likely to generate a 20-plus foot sea-level rise.[65] This rise will occur by 2015, even if carbon emissions are reduced by an unbelievable 80%.[66] Adding merely 1.8° Fahrenheit (1° C) by 2020 would leave between 400 million and 1.7 billion people without water.[67] Another 1.8° C increase by 2050 would result in as many as 2 billion people without water and the extinction of 20–30% of all the world's species.[68] The extreme forecast calls

---

CHANCE FOR CHANGE (2007); CLIMATE CHANGE POLICY: A SURVEY (Stephen H. Schneider, Armin Rosencranz & John O. Niles eds. 2002); CLIMATE CHANGE: SCIENCE, STRATEGIES, & SOLUTIONS (Eileen Claussen, Vicki Arroyo Cochran & Debra P. Davis eds. 2001); EXECUTIVE SUMMARY, SURFACE TEMPERATURE RECONSTRUCTIONS FOR THE LAST 2,000 YEARS, National Academy of Sciences, June 22, 2006, *available at* http://www.nap.edu/catalog.php?record_id=11676; FLANNERY, *supra* note 4; LEIGH GLOVER, POSTMODERN CLIMATE CHANGE 69–136 (2006); HENSON, *supra* note 29; 1 JOHANSEN, *supra* note 10; PITTOCK, *supra* note 36; EU CLIMATE CHANGE POLICY: THE CHALLENGE OF NEW REGULATORY INITIATIVES (Marjan Peeters & Kurt Deketelaere eds. 2006); HOUGHTON, *supra* note 13; INTERGOVERNMENTAL PANEL ON CLIMATE CHANGE, CLIMATE CHANGE 2007, THE PHYSICAL SCIENCE BASIS, SUMMARY FOR POLICY MAKERS (2007), *available at* http://www.ipcc.ch/; MONBIOT, *supra* note 11; J.F. RISCHARD, HIGH NOON: TWENTY GLOBAL PROBLEMS, TWENTY YEARS TO SOLVE THEM 70–75 (2002); NICHOLAS STERN, STERN REVIEW: THE ECONOMICS OF CLIMATE CHANGE 1–67 (2007); SPENCER R. WEART, THE DISCOVERY OF GLOBAL WARMING 160–92 (2003); CHARLES WOHLFORTH, THE WHALE AND THE SUPERCOMPUTER: ON THE NORTHERN FRONT OF CLIMATE CHANGE (2004); John Abatzoglou et al., *A Primer on Global Climate Change and Its Likely Impacts*, in CLIMATE CHANGE: WHAT IT MEANS FOR US, OUR CHILDREN, AND OUR GRANDCHILDREN 11 (Joseph F.C. DiMento & Pamela Doughman eds. 2007); Stephanie B. Ohshita, *The Scientific and International Context for Climate Change Initiatives*, 42 U.S.F.L. REV. 1 (2007); Mary Christina Wood, *Nature's Trust: A Legal, Political, and Moral Frame for Global Warming*, 34 B.C. ENVTL. AFF. L. REV. 577 (2007) (blaming environmental law for America's failure to address climate change).

63. PITTOCK, *supra* note 36 at 64–83, 91 (most consequences highly likely at 91).
64. Associated Press, *Climate Report: 'Highway to Extinction:' Dire Predictions Includes Loss of Species, Increasing Scarcity of Water*, MSNBC.COM, April, 1, 2007.
65. ROMM, HELL AND HIGH WATER, *supra* note 10 at 21–22, 74–76.
66. *Id.*
67. Associated Press, *supra* note 64. *See also* Associated Press, *supra* note 30 (based on draft of climate report predicting tens of millions flooded out of their homes each year and hundreds of millions of Africans and tens of millions in Latin America will be short of water).
68. Associated Press, *supra* note 64. *See also* Associated Press, *supra* note 30 (based on draft of climate report predicting 1 billion in Asia facing water shortages).

for a 7–9° C average temperature increase that would affect 20% of the world population with flooding and 1.1–3.2 billion people with water scarcity.[69]

One of the most influential scientists of the last century, James Lovelock, predicts that by 2040 the Sahara Desert will jump the Mediterranean and expand into Europe. Berlin will be as hot as Baghdad; Atlanta will be a kudzu jungle; and 6 billion people will perish, leaving 500 million survivors residing in Canada, Iceland, Scandinavia, and the Arctic Basin.[70] Lovelock points to the record from 3 million years ago when average temperatures rose by 5° C causing the seas to rise not 23 inches but more than 80 feet—an increase which may be reached by 2030 when the Arctic is projected to be ice-free.[71] He argues that sustainability and renewable energy will make little difference and that the only strategy remaining is "sustainable retreat."[72] The last time the Earth was 1° C warmer than it is today, sea levels were 20 feet higher; the last time the Earth was 2°–3° C warmer, sea levels were more than 80 feet higher.[73] Should the western part of the Antarctic or all of the Greenland ice sheet melt as is feared, the sea rise is projected to be 200 feet.[74] The rate of melting of the Greenland ice cap has accelerated sufficiently to trigger earthquakes as pieces of several cubic kilometers break off, greatly outpacing the predictions of the Intergovernmental Panel on Climate Change (IPCC).[75] The Ilulissat Ice Fjord Glacier in Greenland is flowing three times faster today than it was 10 years ago.[76] Currently, the glacier is flowing two meters per hour on a front five kilometers wide and 1,500 meters deep, putting enough fresh water into the sea each day to provide drinking water to New York City for a year.[77] The rate of flow surges at times

---

69. Associated Press, *supra* note 64. *See also* Associated Press, *supra* note 30 (based on draft of climate report predicting hundreds of millions facing starvation, 1.1 billion to 3.2 billion facing water shortages and 100 million flooded annually by 2080).

70. Jeff Goodell, *The Prophet of Climate Change: James Lovelock*, ROLLING STONE.COM, Oct. 17, 2007, available at http://www.rollingstone.com/politics/story/16956300/the_prophet_of_climate_change_james_lovelock. *See also* JAMES LOVELOCK, GAIA: A NEW LOOK AT LIFE ON EARTH (2000); JAMES LOVELOCK, HEALING GAIA: PRACTICAL MEDICINE FOR THE PLANET (1991).

71. Goodell, *The Prophet of Climate Change, supra* note 70.

72. *Id.*

73. ROMM, HELL AND HIGH WATER, *supra* note 10 at 20.

74. PETER D. WARD, UNDER A GREEN SKY: GLOBAL WARMING, THE MASS EXTINCTIONS OF THE PAST AND WHAT THEY CAN TELL US ABOUT OUR FUTURE 179 (2007).

75. Paul Brown, *Ice Caps Melting Fast: Say Goodbye to the Big Apple?*, ALTERNET, Oct. 15, 2007, available at http://www.alternet.org/story/64735/.

76. *Id.*

77. *Id.*

to five kilometers in 90 minutes.[78] Should Greenland melt, the world sea level would rise seven meters not considering glaciers in Alaska, the Arctic and Antarctica; high tides and hurricane storm surges would cause cities and land all along the coasts to be submerged.[79] Even a one meter sea level rise would devastate 12,000 miles of U.S. coastline where 53% of the nation's population resides, flooding Miami Beach and New Orleans as well as small coastal towns. A 1.5 meter rise would flood Miami, portions of Boston, and Atlantic City; and a 3 meter rise would flood San Francisco, New York, Boston, and Savannah.[80] The seas will likely rise by a foot or two by 2050.[81] Even without a dramatic sea level rise, salt water flooding from storm surges could pollute drinking water in large coastal cities and could cause population to abandon cities such as Shanghai, Manila, Jakarta, Bangkok, Kolkata, Mumbai, Karachi, Lagos, Buenos Aires, and Lima.[82] The Group of Eight Presidencies (G8) representing Europe is proposing to allow a 2° C increase before it takes action to significantly reduce greenhouse gas emissions; President Bush is proposing simply the identification of long-term goals by 2008.[83] In June of 2007, the G8 was unable to commit to even the 2° C cap suggesting that temperatures could rise by 3°–5° C before aggressive action would possibly be undertaken and catastrophe threatens.[84] The failure to act decisively brings the planet closer to the point when stabilization is impossible and conditions unliveable. The latest report from the United Nations, however, warns that "the planet is in danger of crossing a 'tipping point' of irreversible damage to its atmosphere, climate, water, and ecosystems unless governments can develop comprehensive strategies to promote growth and sustainability."[85] Absent a reduction of 90% of carbon emissions, the most conservative of estimates projects a 2° C temperature rise is anticipated by 2030.[86] Climate change models predict that by the 2050s, summer temperatures will have risen by 2.12°–2.75° C resulting in the num-

---

78. *Id.*
79. *Id.*
80. Neal Peirce, *Study Shows High Sea Rise Danger for U.S. Coastal Cities*, POSTWRITERSGROUP.COM, Sept. 9, 2007, available at http://www.postwritersgroup.com/archives/peir070909.htm.
81. Jonathan Barnett & Kristina Hill, *Design for Rising Sea Levels*, HARVARD DESIGN MAGAZINE (No. 27 Fall2007/Winter 2008), *available at* http://www.gsd.harvard.edu/research/publications/hdm/current/27_BarnettHill.html.
82. MONBIOT, *supra* note 11 at 8.
83. *In Hot Seat, Bush Unveils New Climate Strategy*, MSNBC.COM, Jun. 1, 2007.
84. *Id.*
85. GEO_4, *supra* note 19.
86. MONBIOT, *supra* note 11 at 15–19.

ber of heat wave days rising from the average of 14 days annually in London and New York to roughly 50 days per year.[87] Current estimates forecast a increasing average temperatures of 1.5°–4.5° C (2.7°–8.1° F) by the end of the century.[88] The Kyoto Protocol called for only a 5.2% reduction of carbon emissions by 2012.[89] Other estimates exist that carbon emissions could even double generating a temperature rise of 11.5° C by the end of the century[90]

At this time, habitats have been permanently altered, oceans acidified, wetlands lost, coral reefs bleached, and allergy-inducing pollen increased.[91] In Alaska, the town of Newtok has become an island and will soon be under water along with 180 villages that are eroding and sinking as the permafrost melts.[92] The risk of catastrophic global climate disruption could be mitigated, however, if carbon emissions in the 21st century could be limited to 500 billion tons rather than the 1,400 billion tons that are projected.[93] Leadership is in denial on the issue of extinction through climate change yet the risk creates a moral argument for aggressive action.[94]

---

87. KAHN, *supra* note 25 at 133.

88. Daniel A. Farber, *Basic Compensation for Victims of Climate Change*, 155 U. PA. L. REV. 1605, 1606 (2007), *citing* Richard A. Kerr et al., *Latest Forecast: Stand By for a Warmer But not Scourching, World*, 312 SCIENCE 351, 351 (2006). *See generally* Real Climate, http//www.realclimate.org (last visited May 1, 2007) (up to date climate science).

89. MONBIOT, *supra* note 11 at 48.

90. *Id* at 6.

91. 3 JOHANSEN, *supra* note 1; Wagner, *supra* note 1; Associated Press, *supra* note 30 (based on draft of climate report).

92. William Yardley, *Victim of Climate Change, a Town Seeks a Lifeline*, N.Y. TIMES, May 27, 2007.

93. Steven Bernow, *et al.*, *Carbon Abatement with Economic Growth: A National Strategy, in* CLIMATE CHANGE POLICY: A SURVEY 189, 189 (Stephen H. Schneider, Armin Rosencranz & John O. Niles eds. 2002).

94. GORE, *supra* note 4. *See also* DONALD BROWN, AMERICAN HEAT: ETHICAL PROBLEMS WITH THE UNITED STATES RESPONSE TO GLOBAL WARMING (2002); Dale Jamieson, *Ethics, Public Policy and Global Warming* in MORALITY'S PROGRESS 282 (2003).

Chapter 3

Agriculture and Food Policy

> Every year, because of our misuse of the Earth's resources, we lose 100 million acres of farmland and 24 billion tons of topsoil, and we create 15 million acres of new desert around the world. An inch of good topsoil can take a thousand years to form, but when people destroy windbreaks by cutting down trees, the topsoil can be washed or blown away in months.
>
> James Martin, The Meaning of the 21st Century 4 (2006)

Agricultural, and more specifically food policies, are intricately connected to carbon emissions.[95] First, food supplies are often transported over long distances, even between continents, to supply today's consumers. Transporting fish or other fresh products thousands of miles by airplane to markets produces extraordinary emissions. Similarly, long trucking hauls also emit significant carbon. Second, farming methods, including the use of farm equipment, irrigation, fertilizers, and other supplies, may utilize a significant amount of power and generate emissions. Modifications to agricultural methods can reduce emissions and produce greater efficiency and sustainability. Erecting a wind generator can cut the farmer's dependency on carbon-generating electricity.[96] Double cropping by growing soy beans in summer followed by winter wheat is beneficial as soy beans fix nitrogen in the soil reducing the need to apply fertilizer and thus the emissions associated with its production and transportation.[97]

---

95. *See generally* Richard Manning, Against the Grain: How Agriculture Has Hijacked Civilization (2004).

96. Henry C. Jackson, *Farmer Take Another Look at Wind Energy*, Miami Herald, Sept. 24, 2007, *available at* http://www.miamiherald.com/business/AP/story/248358.html (reporting that a farmer in Iowa constructed a wind turbine for $140,000 less a federal government grant of $29,000 that produces twice the energy used and could sell the excess or bank it with the power company for when there is no wind with an investment payoff of 10 to 15 years which shorten should electricity rates increase as anticipated while generating no carbon emissions).

97. Brown, *supra* note 36 at 165.

A quite typical meal might include strawberries from Costa Rica, asparagus from South Africa, chicken from Thailand, Spanish carrots, Zimbabwe snowpeas, and Italian potatoes that may have cumulatively traveled more than 24,000 miles to reach the dinner table.[98] To reduce the distance that food travels, its production must become more local. After peak oil,[99] when supplies are on the decline, air transportation for food and produce will be prohibitive.[100] Oil exhaustion, however, is unreachable for, if remaining fossil fuel reserves are burned, the carbon emissions would equal 18 times the carbon generated during the past 250 years probably yielding the extinction of the population.[101] This calls for community gardens and urban agriculture.[102] American restaurants and consumers must shift to local, in-season products, and cease combing the world to maintain the identical menus year-round. Consumers must demand disclosure of the geographical source of products purchased and consumed and insist on local and regional products. The future of agriculture may call for urban high-rise farms that involve green architecture.[103] Alternatively, hydroponic gardens, greenhouses, canning, smoking, and other storage methods will be required in climates with long winters or droughts.

The type of food consumed also impacts carbon emissions. The consumption of beef requires extensive cultivation of feed, necessitating more farming operations involving trucks, energy, and production. The rapid expansion of global population calls for more food production and the demand for beef

---

98. Peter Singer & Jim Mason, The Way We Eat: Why Our Choices Matter 136 (2006).

99. James Howard Kunstler, The Long Emergency: Surviving the Converging Catastrophes of the Twenty-First Century 24–28 (2005); Greg Pahl, The Citizen-Powered Energy Handbook: Community Solutions to a Global Crisis ix–xxv (2007); Peter Tertzakian, A Thousand Barrels a Second: The Coming Oil Break Point and the Challenges Facing an Energy Dependent World (2006).

100. Brown, *supra* note 36 at 39.

101. Mayer Hillman et al, The Suicidal Planet: How to Prevent Global Climate Catastrophe 69 (2007).

102. *See generally* Global Development of Organic Agriculture: Challenges and Prospects (Niels Halberg et al. eds. 2006).

103. Lisa Chamberlain, *Skyfarming*, N.Y. Magazine, Apr. 9, 2007, *available at* http://www.nymag.com/news/features/30020 (discussing the research of Dickson Despommier of Columbia University); Gretchen Vogel, *Upending the Traditional Farm: Cities are Taking Over farmland. Could They Someday Take Over the Job of Farming, Too,"* 319 Science 752 (Feb. 2008). *See also Vertical Farming*, Wikipedia, *available at* http://en.wikipedia.org/wiki/Vertical_farming.

Cattle Feed Yard. Source: USDA.

causes more deforestation to develop agricultural fields to feed the ever-increasing number of livestock.[104] The loss of forest land reduces carbon sequestration and the conversion of carbon dioxide to oxygen through photosynthesis. A field of cows produces less than a tenth of the nutrients of a field of vegetables.[105]

Sequestration refers to natural and man-made systems that consume or trap carbon emissions and thereby prevent them from entering the atmosphere. A form of sequestration is the "carbon sink," a natural or a man-made system or planting of crops that trap carbon emissions. Carbon sequestration through forestation, expansion of green space, or land-based "carbon sinks" such as farms to capture carbon may be useful but forests, in addition to sequestering carbon, absorb more radiation and heat.[106] Some argue that agricultural lands

---

104. Andrew Manale, *Agriculture and the Developing World: Intensive Animal Production, a Growing Environmental Problem*, 19 GEO. INT'L ENVTL. L. REV. 809 (2007) (economic development from developing nations supplying animals but import of grains costly and environmental degradation a by-product).
105. MARTIN, THE MEANING OF THE 21ST CENTURY, *supra* note 32 at 74.
106. PITTOCK, *supra* note 36 at 183–88.

must be returned to forest and agriculture shifted to vertical urban farms to slow and reverse global warming.[107] Forestation, however, is complicated as forests naturally burn and the released carbon dioxide may overshadow the sequestration properties.[108] There also exists some evidence that tree planting and forestation through increased fire danger and reduced sequestration capacity may actually increase carbon emission.[109] Canadian forests prior to 1970 absorbed 118 million tons of carbon annually, which exceeded that nation's emissions.[110] However, by 1995, forests released 57 million tons of carbon annually, becoming an emitter rather than an absorber of carbon dioxide.[111] Old forests have greater carbon sequestration capacity than do newer growth forests.[112] Reforestation can carry unanticipated consequences, as in Japan where the largest reforestation program in the world has been underway and has spawned an extraordinary infestation of bloodsucking leeches.[113]

Some farmers are producing carbon sequestration through foresting former agricultural land to profit from the sale of carbon credits[114] which are rapidly appreciating in Europe.[115] Credits are part of programs for trading emissions credits such as under the Kyoto Protocol and some state pollution control programs under which industry and other polluters or nations are allotted a limited number of credits or subjected to a reduced level of emissions. Polluters or polluting nations are then free to choose whether to reduce emissions or to purchase credits from those who possess excess credits, such as those developing carbon sinks or sequestration capacity, or from entities who are reducing emissions, such as a farmer that plants trees. Similarly, non-polluting or exempt poorer nations can sell their excess polluting capacity credits to devel-

---

107. Chamberlain, *supra* note 103 (discussing the research of Dickson Despommier of Columbia University); Vogel, *supra* note 103. *See also Vertical Farming*, WIKIPEDIA, available at http://en.wikipedia.org/wiki/Vertical_farming.

108. PETER F. SMITH, ARCHITECTURE IN A CLIMATE OF CHANGE: A GUIDE FOR SUSTAINABLE DESIGN 15 (2001)

109. 1 JOHANSEN, *supra* note 10 at 37–41.

110. GELBSPAN, THE HEAT IS ON, *supra* note 11 at 21.

111. *Id.*

112. 3 JOHANSEN, *supra* note 1 at 660–61.

113. Andrew Leonard, *A Plague of Bloodsuckers: In Japan, Reforestation, Population Decline and Global Warming Have Set Off a Land Leech Invasion*, SALON.COM, Sept. 7, 2007, available at http://www.salon.com/tech/htww/2007/09/07/a_plague_leeches/print.html.

114. *Sale of Carbon Credits Helping Land-Rich, but Cash-Poor, Tribes*, N.Y. TIMES.COM, May 7, 2007 (reporting Indian tribal activities with credits at $4 per ton and $30 per ton in Europe and anticipated to constitute sufficient incentive at $12).

115. *Id.*

oped polluting nations. Overall, the carbon trading system might support a reduction of cumulative emissions.[116]

As oil and gasoline become scarce and more expensive, the drive to develop ethanol from crops increases. This is a development that will be catastrophic because 800 million cars will be competing for food resources with 1.2 billion people living on less than $1 a day.[117] Jean Ziegler, the United Nations' independent expert on the right to food called the growing practice of converting food crops into biofuel "a crime against humanity," saying it is creating food shortages and price jumps that cause millions of poor people to go hungry.[118] Most of the 100 million tons of cereal grain which is nearly all corn ammounting to 12% of the corn consumed on the planet has gone to biofuel resulting in rising prices of food and a 5% decline in food stocks.[119] Due to the doubling of corn prices resulting from ethanol production competiing with food pro-

---

116. Erik B. Bluemel, *Unraveling the Global Warming Regime Complex: Competitive Entropy in the Regulation of the Global Public Good*, 155 U. PA. L. REV. 1981 (2007) (advocating worldwide trading and a clean development fund); Christopher Carr & Flavia Rosembuj, *Flexible Mechanisms for Climate Change Compliance: Emission Offset Purchases Under the Clean Development Mechanism*, 16 N.Y.U. ENVTL. L.J. 44 (2008) (endorsing carbon credit trading); B. Timothy Heinmiller, *The Politics of "Cap and Trade" Policies*. 47 NAT. RESOURCES J. 445 (2007); George (Rock) Pring, *A Decade of Emissions Trading in the USA: Experiences and Observations for the EU*, in EU CLIMATE CHANGE POLICY: THE CHALLENGE OF NEW REGULATORY INITIATIVES 188–201 (Marjan Peeters & Kurt Deketelaere eds. 2006). *See also* Dennis Hirsch et al., *Emissions Trading—Practical Aspects*, in GLOBAL CLIMATE CHANGE AND U.S. LAW 627 (Michael B. Gerrard ed. 2007); Nick Johnstone, *Tradable Permits for Climate Change: Implications for Compliance, Monitoring, and Enforcement*, in CLIMATE-CHANGE POLICY 238 (Dieter Helm ed. 2005); Steven Sorrell & Jos Sijm, *Carbon Trading in the Policy* Mix, in CLIMATE-CHANGE POLICY 194 (Dieter Helm ed. 2005); Jonathan Donehower, Comment, *Analyzing Carbon Emissions Trading: A Potential Cost Efficient Mechanism to Reduce Carbon Emissions*, 38 ENVTL. L. 177 (2008) (arguing that without the incorporation of China and the United States, the world's two largest polluters, the carbon markets may serve as a successful market tool and example of the efficiency of an open market to cost-efficiently solve environmental problems, but will do nothing to curb GHG emissions and limit the effects of climate change).

117. BROWN, *supra* note 36 at 8, 25–36. *But see* Siwa Msangi & Mark Rosegrant, *Agriculture and the Environment: Linkages, Trade-offs and Opportunities*, 19 GEO. INT'L ENVTL. L. REV. 699 (2007) (suggesting that biofuel development carries economic opportunities that may outweigh food versus fuel debates with perhaps excessive optimism for environmental impacts and problems of water and emissions).

118. Edith M. Lederer, *UN Expert Decries Turning Food Into Fuel*, YAHOO NEWS, Oct. 27, 2007, *available at* http://news.yahoo.com/s/ap/20071026/ap_on_re_us/un_food_vs_biofuel_3.

119. Edith M. Lederer, *Biofuel Growth Adds to Hunger: Most Vulnerable 'Priced Out' of Market for Food*, Wash. Times, Feb. 14, 2008.

ducers, record levels of corn have been produced since 2002.[120] The enlarged corn crop required extremely high levels of fertilization that then generated higher levels of nitrogen that drained into the Mississippi River.[121] The result has been a 7,900 square mile dead zone in the Gulf of Mexico that suffocates fish, crabs and shrimp.[122] Demand for biofuels has resulted in agricultural land dedicated to that use increasing from 30 million acres in 2002 to 200 million acres in 2008.[123] Biofuel currently supplies 3% of global energy needs and is anticipated to rise to 10.6% by 2030.[124]

Controversy exists over the impacts of greenhouse gases on agriculture. While some believe that production in North America and Europe will result in higher yields, evidence exists that increasing carbon dioxide may destroy soil nitrogen.[125] This would seriously limit growth and plants will cease to act as a carbon sink sequestering carbon dioxide as they have in the past.[126] The quality of plant growth may also decline with increased carbon dioxide in the atmosphere.[127] Higher temperatures can reduce or even halt photosynthesis, prevent pollination, and lead to crop dehydration.[128] The detrimental effect of higher temperatures on yields overrides the carbon dioxide fertilization effect.[129]

Worldwide, 70% of all water consumed is used for irrigation.[130] More efficient crops should be substituted for those that demand more water, e.g., rice that yields 4 tons per acre uses only a little more water than rice that yields 2 tons an acre,[131] and wheat which typically produces 50% more calories per unit of water than rice, and most protein crops, can be modified to produce a high yield with less water.[132] Replacing rotating sprinklers with drip irriga-

---

120. Associated Press, *Corn Boom Could Expand 'Dead Zone' in Gulf*, MSNBC, Dec. 17, 2007, *available at* http://www.msnbc.msn.com/id/22301669/.
121. *Id.*
122. *Id.*
123. Ambrose Evans-Pritchard, *Why the Price of 'Peak Oil' is Famine*, TELEGRAPH.CO.UK, Feb. 9, 2008, *available at* http://www.telegraph.co.uk/money/main.jhtml?xml=/money/2008/02/07/cnoil107.xml.
124. *Id.*
125. 1 JOHANSEN, *supra* note 10 at 41–43.
126. *Id.*
127. 3 JOHANSEN, *supra* note 1 at 547–556.
128. BROWN, *supra* note 36 at 64.
129. *Id.*
130. MARTIN, THE MEANING OF THE 21ST CENTURY, *supra* note 32 at 70.
131. *Id.*
132. *Id.*

tion saves a great deal of water and electronic systems can reduce water use by 70%.[133]

Urbanization, drought, excessive water and irrigation use, and poor agricultural practices can cause the loss of prime topsoil. Most soil erosion is caused by water, either through flooding or poor irrigation, with the remainder lost to winds.[134] Ploughing and repeated planting, which depletes soil of nutrients, also destroys farmland soil.[135] Each square yard of top soil contains thousands of spiders, ants, and wood lice, beetles, and fly larvae, 2,000 earthworms, 20,000 pot worms, 2,000 millipedes and centipedes, 8,000 slugs, 40,000 springtails—minute primitive insects, 120,000 mites, and 12 million nematodes—or roundworms.[136] Each gram of topsoil contains 4,000 distinct genomes, differing greatly in different locations.[137] One teaspoon of good grassland soil may contain 5 billion bacteria, 20 million fungi and a million protoctists—a diverse group of species, ranging from single-celled organisms to multi-cellular organisms such as seaweed and kelp; grouped together mainly because they cannot be classified as plants, animals, fungi or bacteria[138] To illustrate the density of life forms below the ground by weight, they are the equivalent of 12 horses per acre.[139] The complex soil is being lost to winds and urbanization destroyed in farmlands by the use of herbicides, pesticides, and crude fertilizers.[140]

One-third of America's original topsoil is gone and the rest is degraded.[141] Ninety percent of farmland in the United States is losing topsoil an average of 17 times faster than new topsoil is being formed.[142] In much of the rest of the world, soil loss per ton of food produced exceeds the loss occurring in the United States.[143] Forty percent of the world's agricultural land is degraded.[144] In Central America, 75% of the land is infertile, in Africa 20%, and in Asia

---

133. *Id.*
134. Ian Sample, *Global Food Crisis Looms as Climate Change and Population Growth Strip Fertile Land*, GUARDIAN, Aug. 31, 2007, *available at* http://www.guardian.co.uk/environment/2007/aug/31/climatechange.food.
135. *Id.*
136. MARTIN, THE MEANING OF THE 21ST CENTURY, *supra* note 32 at 72.
137. *Id.*
138. *Id* at 71.
139. *Id* at 72.
140. *Id.*
141. *Id.*
142. *Id.*
143. *Id.*
144. Sample, *supra* note 134.

Poultry Production. Source: USDA/Larry Rana.

11% is unsuitable for farming due to soil degradation.[145] Forty percent of the world's harvest is lost in the fields: 13% is lost to disease, 15% destroyed by insects, 12% by weeds and 10% from other causes.[146] These problems can be improved through multi-cropping or changing crops each year to three years such as corn to soybeans, but U.S. government subsidies discourage the practice.[147] Yet, as much as one-third of all U.S. agricultural crops are lost annually to petroleum-related air and water pollution.[148]

Unfortunately, due to the political influence of agri-business, "[m]odern trends and theories in agriculture are depopulating the countryside, spoiling the land, squandering the water, poisoning the food, deepening the global divisions between rich and poor, and threatening whole ecosystems.[149] At the extreme of projections, albeit possibly unrealistic, dense cities will leave no

---

145. *Id.*
146. Martin, The Meaning of the 21st Century, *supra* note 32 at 73.
147. *Id.*
148. Terry Tamminen, Lives Per Gallon: The True Cost of Our Oil Addiction 207 (2006).
149. George Pyle, Raising Less Corn, More Hell: The Case for the Independent Farm and Against Industrial Food xx (2005).

room for gardens, and heat will preclude agriculture so that the future will require synthesized food.[150] It is estimated that without cheap fossil fuels the United States will require 50 million new farmers, one for every two households or 25 times more than the current number of farmers.[151] The nation's 30 million acres of lawns, however, can be converted to crops and could yield three times the current land planted for irrigated corn.[152]

1. Local Response

To reduce emissions, local laws can be amended to permit and encourage local agriculture through amending zoning to make urban farms permissible (subject to site plan approval to assure mitigation of nuisances). Farming methods can be regulated to require fewer emissions, such as through ordinances requiring soil and woodland regeneration, prohibiting till farming (as plowing generates carbon dioxide emissions) or providing for reduced furrows, cover crops, nutrient management, manuring and sludge application, improved grazing, water conservation, efficient irrigation, agroforestry practices, and requiring the growth of energy crops on spare land.[153] The elimination of pesticides and herbicides can lower emissions by reducing production demand for them and their associated transportation and can also protect valuable water tables.[154] The purchase or establishment of local farms by government or local investors or farmers to advance programs for "buy locally grown" strategies can support farming. Through farmers' markets and local supply fresh, local, non-genetically altered produce can be made available to an expanding urban market.[155] Some municipally-sponsored farmers markets, as New York City's 47 greenmarkets, already limit participation to local farmers; nationwide 19,000 farmers sell only at local farmers' markets.[156] In addition, organic roofs can be required in construction and density bonuses or tax credits might be offered for the development of roof gardens and converting lawns to food

---

150. Goodell, *The Prophet of Climate Change, supra* note 70.
151. Wylie Harris, *Lawn to Farm: Suburbia's Silver Lining*, COMMON DREAMS.ORG, Jan. 24, 2008.
152. *Id.*
153. 3 JOHANSEN, *supra* note 1 at 666–67.
154. *See generally* RACHEL CARSON, SILENT SPRING (1962).
155. Rick Wartzman, *Can the City Save the Farm?*, CAL. MAGAZINE, May/June, 2007, *available at* http://www.newamerica.net/publications/articles/2007/can_city_save_farm_5422.
156. SINGER & MASON, *supra* note 98 at 137–38.

cultivation. Ordinances can be adopted restricting garden and agricultural products and encouraging carbon sequestration, food, irrigation conservation, and soil protection. Surface water collection for irrigation can also be required. Incentives such as tax benefits and offering transferable development rights should be used to encourage reforestation.

Agriculture has a significant impact on global warming but that impact can be mitigated through sequestration and the reduction of emissions, including the reduction of burning, deforestation and land degradation.[157] Under a tradable carbon permit scheme, farmers could grow a crop of carbon sequestering produce in exchange for the value of permits from firms that need to reduce emissions.[158] Lowering the nitrogen content of fertilizer is possible while maintaining crop yield and increasing profitability.[159] Fertilization and irrigation are both energy intensive.[160] Wise use of cover crops, manure, and utilizing manure digesters can also reduce greenhouse gasses.[161] Digesters restrict oxygen yielding methane gas and reduce the volume of solids and liquids to be treated.[162] The methane can be sold or used to generate electricity on the farm. The solid matter left behind is a valuable soil amendment and the liquids become an easily applied fertilizer with plant-available nutrients and low pathogen levels. The digesters, although not emission-free, thus present a safer, more efficient, more sustainable form of fertilizer, producing bio-solids and methane which can be burned off or used as green fuel.

Constitutionally, the most controversial proposal would be limiting food markets and restaurants to sell only regionally-produced food items.[163] Al-

---

157. Holly L. Pearson, *Climate Change and Agriculture: Mitigation Options and Potential*, in CLIMATE CHANGE POLICY: A SURVEY 307, 307–09 (Stephen H. Schneider, Armin Rosencranz & John O. Niles eds. 2002).

158. Richard M. Adams *et al.*, *Impacts on the Agricultural Sector*, in CLIMATE CHANGE: SCIENCE, STRATEGIES, & SOLUTIONS 25, 40 (Eileen Claussen, Vicki Arroyo Cochran & Debra P. Davis eds. 2001).

159. Pearson, *in* CLIMATE CHANGE POLICY, *supra* note 157 at 316.

160. *Id* at 316–17.

161. *Id* at 327.

162. U.S. Department of Agriculture, *How Anaerobic Digestion (Methane Recovery) Works*, available at http://www.eere.energy.gov/consumer/your_workplace/farms_ranches/index.cfm/mytopic=30003.

163. *Cf.* Daisuke Kojo, *The Importance of the Geographic Origin of Agricultural Products: A Comparison of Japanese and American Approaches*, 14 MO. ENVTL. L. & POL'Y REV. 275 (2006) (discussing place of origin labeling requirements on perishable food and the stronger Japanese standards)

though the Commerce Clause of the U.S. Constitution[164] would ostensibly impose constraints on regulating access to markets by foreign or interstate products,[165] ordinances that restrict food sold, prepared, or served to products that are regionally grown through "local food" ordinances are needed as a compelling state interest of survival of mankind and should pass constitutional review under the Commerce Clause[166] and the Privileges and Immunities Clause despite discrimination against out-of-state products.[167] Just as cigarettes and alcohol can be separately taxed, so can beef and the meat of other animals to discourage consumption.

2. Regional Response

At the regional level, urbanized development must be halted at the municipal limits and zoning set for forest and essential agriculture. Master plans should be coordinated with plans for increasing urban agriculture. Tax sharing of the regional tax revenues should be undertaken so that rural areas and nonurbanized areas do not have to compete for commercial uses and so that incentives can be offered to adopt sustainable agriculture policies.[168] Where

---

164. U.S. CONST., Art. I, §8.
165. West Lynn Creamery, Inc. v. Healy, 512 U.S. 186 (1994) (invalidating burden on milk importation); City of Philadelphia v. New Jersey, 437 U.S. 617 (1978) (invalidating ban on waste importation); H.P. Hood & Sons, Inc. v. DuMond, 336 U.S. 529 (1939) (right to export goods).
166. Maine v. Taylor, 477 U.S. 131 (1986) (validating ban on bait fish importation as no less restrictive alternative to achieving compelling state interest of protecting natural resources from potential parasites). *See also* United Haulers Ass'n v. Oneida-Herkimer Solid Waste Management Auth., 127 S. Ct. 1786 (2007) (county flow ordinances requiring all solid waste generated within county to be delivered to public owned waste processing facility not violative of commerce clause by forcing haulers to pay more for waste disposal than out-of state facilities, where the flow ordinance required delivery to a private recycler, finding no discrimination and compelling reasons of allocating costs to the citizens and allowing local government to pursue chosen policies). Cf. Peter Carl Nordberg, Note, *Excuse Me, Sir, But Your Climate's on Fire: California's S.B. 1368 and the Dormant Commerce Clause*, 82 NOTRE DAME L. REV. 2067, 2074–75 (2007) (arguing that state legislation limiting purchases of out-of-state energy that fails to meet global warming standards should be sustained under the Commerce Clause).
167. U.S. CONST., Art. IV, §2; Supreme Court of New Hampshire v. Piper, 470 U.S. 274 (1985) (recognizing compelling state interest justification yet finding it lacking in a residency requirement for attorneys).
168. MYRON ORFIELD, AMERICAN METROPOLIS: THE NEW SUBURBAN REALITY 105–108 (2002); DAVID RUSK, INSIDE GAME—OUTSIDE GAME: WINNING STRATEGIES FOR SAVING URBAN

development is thwarted, transferable development rights[169] can be shifted to densification areas such as around regional transit and urban infill areas to allow protection of agricultural and other rural land. Congestion pricing should be imposed, whereby tolls can be imposed for entry into congested areas or on particular roads, with exemptions for essential agricultural and local food delivery, to discourage vehicle use of rural roads and highways.[170] Yet congestion pricing schemes are a small fraction of the subsidy offered by parking.[171] Local rural residents might receive a discount but should be encouraged to relocate to urbanized areas.

3. State Response

Farmers can be offered the opportunity to earn and sell greenhouse gas emissions credits as has been done in Illinois[172] through tree planting and other forms of carbon sinks or sequestration. States can also impose emissions standards on vehicles and farm equipment.[173] State legislatures are also capable of offering a broad assortment of incentives to engage in sustainable agriculture and to stimulate the development of local and urban agriculture. A state might also impose local and regional produce laws just as local communities do.

AMERICA (1999); Note, *Making Mixed-Income Communities Possible: Tax Base and Class Desegregation*, 114 HARV. L. REV. 1575 (2001).

169. 1 JAMES A. KUSHNER, SUBDIVISION LAW AND GROWTH MANAGEMENT § 2:13 (2d ed. 2001 & Supp. 2008).

170. DAVID BANISTER, UNSUSTAINABLE TRANSPORT: CITY TRANSPORT IN THE NEW CENTURY 130–45 (2005); TIMOTHY BEATLEY, GREEN URBANISM: LEARNING FROM EUROPEAN CITIES 151–61 (2000); ROBERT CERVERO, THE TRANSIT METROPOLIS: A GLOBAL INQUIRY (1998); JAMES A. KUSHNER, THE POST-AUTOMOBILE CITY 102–04 (2004); Ian Parry & Elena Safirova, *Pay as You Slow: Road Pricing to Reduce Traffic Congestion, in* NEW APPROACHES ON ENERGY AND THE ENVIRONMENT: POLICY ADVICE FOR THE PRESIDENT 63 (Richard D. Morgenstern & Paul R. Portney eds. 2004).

171. DONALD C. SHOUP, THE HIGH COST OF FREE PARKING 215–17 (2005).

172. *See* U.S. Environmental Protection Agency, Global Warming, Actions, State, http://yosemite.epa.gov/OAR/globalwarming.nsf/content/ActionsState.html (last visited Sept. 19, 2006). *See generally* Kirsten Engel, *State and Local Climate Change Initiatives: What is Motivating State and Local Governments to Address a Global Problem and What Does This Say About Federalism and Environmental Law?*, 38 URB. LAW. 1015 (2006).

173. *See* Central Valley Chrysler Valley Jeep, Inc. v. Witherspoon, No.CV-F-04-6663 AWI LJO, 2006 WL 2473663, at 1 (E.D. Cal. Aug. 25, 2006). *See generally* Ann Carlson & Tim Malloy, *Special Edition: California's AB 1493: Trendsetting or Setting Ourselves Up To Fail?*, 21 UCLA J. ENVTL. L. & POL'Y 97, 102 (2003).

These can take the form of regulation limiting food travel distance or imposing excise taxes on transported items.

4. Federal Response

To avoid judicial challenges to state and local regulation of food, Congress would be in the best position to enact legislation to advance sustainable agricultural and food policies. Policies might include capping the distance that food may travel to assure a local and regional food supply, establishing obligations for investment in renewable energy, reducing energy consumption and emissions, prohibiting fertilizer use, shifting from other unsustainable farm practices, and modifying tax incentives to reward carbon sequestration and energy reduction investments through tax credits. Congress should eliminate existing farm subsidy programs and prohibit foreign food shipments beyond a set number of miles unless the shipment is deemed humanitarian in nature. This should provide support for indigenous agriculture in poorer countries, advance "local food" policies, and encourage economic development. If subsidies are to be utilized, they should target fresh local produce for salads and raw eating with taxes imposed on non-regional food, meat, high-fat foods, and food requiring extensive cooking. Congress should also increase regulation to assure safe food and safe food production. Congress should subsidize healthy foods, foods that require little or no transportation and little or no energy to cook. Foreign aid should be targeted at programs to develop local agriculture and food production.

The draining of water tables and surface water will significantly escalate with global warming and population growth and will seriously reduce the supply of food.[174] It is imperative to conserve water and institute efficient irrigation and agricultural practices.

5. International Response

The conservation of open space and forests must be supported and aid for forestation projects made available.[175] Unfortunately, politics has hampered agreements to restrict deforestation and the prospects for improvement are dim.[176] Afforestation, the planting of forests, however, would provide signifi-

---

174. BROWN, *supra* note 36 at 57–58.
175. Niles, *in* CLIMATE CHANGE POLICY, *supra* note 45 at 355.
176. *Id* at 360–65.

cant mitigation.[177] Internationally, standards could be imposed to reduce energy use in food production and agriculture and to impose local and regional food source regulations.

The United States, the European Union, Canada, Australia, and other rich countries provide $350 billion a year in subsidies to farmers.[178] That is seven times what those nations spend on foreign aid to poorer nations.[179] The $350 billion in agricultural subsidies is part of a larger list of perverse subsidies amounting to $2 trillion.[180] The subsidies cost the American family $2,000 annually and could supply sufficient resources to balance the budget, eliminate the deficit, and increase health and education spending by 50%.[181] When subsidies encourage over-production and farmers are permitted to profitably dump produce at below cost prices in poor nations, the local agricultural industry is destroyed often sending impoverished farmers to the cities to search for subsistence.[182] Forty percent of a European Union farmer's income comes from subsidies as does 23% of the American farmer's income.[183]

Many of the recommendations for federal action also apply at the international level. So again, transport should be limited in kilometers or miles with the exception of transport by shipping for humanitarian reasons only through treaties. An international carbon emissions trading system might allow farmers with carbon credits accumulated by increasing forestation and carbon sequestration to sell the credits to non-complying industries or nations.[184] As of

---

177. HOUGHTON, *supra* note 13 at 177–79.
178. PYLE, *supra* note 149 at 49.
179. *Id.*
180. MARTIN, THE MEANING OF THE 21ST CENTURY, *supra* note 32 at 46–47.
181. *Id* at 47.
182. PYLE, *supra* note 149 at 49–54.
183. *Id* at 50. *See also* Daniel Bianchi, *Cross Compliance: The New Frontier in Granting Subsidies to the Agricultural Sector in the European Union*, 19 GEO. INT'L ENVTL. L. REV. 817 (2007) (under reformed scheme, 40% of EU budget for agricultural subsidies).
184. *Cf.* Bluemel, *supra* note 116 (advocating worldwide trading and a clean development fund); Jan-Tjeerd Boom & Andries Nentjes, *Alternative Design Options for Emissions Trading: A Survey and Assessment of the Literature*, in CLIMATE CHANGE AND THE KYOTO PROTOCOL: THE ROLE OF INSTITUTIONS AND INSTRUMENTS TO CONTROL GLOBAL CHANGE 45 (Michael Faure *et al.* eds. 2003); Johnstone, in CLIMATE-CHANGE POLICY, *supra* note 116 at 238; Sorrell & Sijm, *in* CLIMATE-CHANGE POLICY, *supra* note 116 at 194; Rethinking *the Kyoto Protocol: Are There Legal Solutions to Global Warming and Climate Change?*, 5 WASH. U. GLOBAL STUD. L. REV. 333 (2006) (Moderator: Douglas Williams; participants: Anita Halvorssen, J. Kevin Healy, William Pizer and Jacob Werksman); Donehower, *supra* note 116 (arguing that without the incorporation of China and the United States, the world's two largest polluters, the carbon markets may serve as a successful market tool and exam-

July 28, 2006, carbon dioxide credits could be purchased for $4.50 per ton in the Chicago Climate Exchange.[185] An international prohibition on deforestation should be coupled with aid to any developing country that is adversely affected economically.

---

ple of the efficiency of an open market to cost-efficiently solve environmental problems, but will do nothing to curb GHG emissions and limit the effects of climate change); Jennifer P. Morgan, Note, *Carbon Trading Under the Kyoto Protocol: Risks and Opportunities for Investors*, 18 FORDHAM ENVTL. L. REV. 151 (2006).

185. John C. Dernbach, *U.S. Policy*, in GLOBAL CLIMATE CHANGE AND U.S. LAW 61, 82 (Michael B. Gerrard ed. 2007).

Chapter 4

Brownfield Development

> Just as in Europe, brownfield redevelopment will become more attractive to real estate developers and investors in the United States when several events occur: when the law ceases to impose unanticipated liability; when the government offers sufficient incentives to stimulate brownfield cleanup and redevelopment; and when responsible smart growth policies are imposed to target real estate development to urban infill and brownfield rejuvenation.
>
> James A. Kushner, *Brownfield Redevelopment Strategies in the United States*, 22 Ga. St. U. L. Rev. 857, 875 (2006)

Brownfields are lands that have been polluted and have been passed over for new development.[186] These are former industrial sites, military bases, rail yards, filling stations, docklands, and other property that contains soil or water pollution. A principal strategy in pursuit of sustainability is to develop compact communities, utilizing infill development rather than expanding urbanization borders into open space, forests, or agricultural districts. European brownfield

---

186. James A. Kushner, *Brownfield Redevelopment Strategies in the United States*, 22 Ga. St. U. L. Rev. 857 (2006). *See also* Charles Bartsch & Elizabeth Collaton, Brownfields: Cleaning and Reusing Contaminated Properties 2–3 (1997); International City/County Management Ass'n, Measuring Success in Brownfields Redevelopment Programs (2002); Mark Reisch & David M. Bearden, Superfund and the Brownfields Issue (2003); Joel B. Eisen, *"Brownfields of Dreams"?: Challenges and Limits of Voluntary Cleanup Programs and Incentives*, 1996 U. Ill. L. Rev. 883, 890–91 (1996); Jennifer Felten, *Brownfield Redevelopment 1995–2005: An Environmental Justice Success Story?* 40 Real Prop. Prob. & Tr. J. 679 (2006) (supporting brownfield redevelopment and arguing that projects are more successful with community involvement); Denice Ferkick Hoffman & Barbara Coler, *Brownfields and the California Department of Toxic Substances Control: Key Programs and Challenges*, 31 Golden Gate U. L. Rev. 433 (2001); Bradford C. Mank, *Reforming State Brownfield Programs to Comply with Title VI*, 24 Harv. Envtl. L. Rev. 115, 120 (2000); Richard G. Opper, *The Brownfield Manifesto*, 37 Urb. Law. 163 (2005); Hope Whitney, *Cities and Superfund: Encouraging Brownfield Redevelopment*, 30 Ecology L.Q. 59 (2003).

Millennium Housing, dockyards to wetlands and housing, London.
Source: James A. Kushner.

development has been extraordinary, allowing urban densification, economic development and regeneration without urban sprawl and generating renewal of entire regions.[187] Brownfield development is an essential policy that regenerates communities, enhances the tax base with new jobs and land values as well as offering an alternative to urban sprawl development. Various funding sources are currently available for brownfield cleanup and redevelopment.[188] Creating more compact, higher density communities connected by public transport will reduce carbon emissions, particularly through reducing automobile emissions and reduce the community's carbon and ecological footprint.

---

187. James A. Kushner, *Brownfield Redevelopment Strategies in the United States*, 22 GA. ST. U. L. REV. 857, 871–73 (2006); James A. Kushner, *Social Sustainability: Planning for Growth in Distressed Places—the German Experience in Berlin, Wittenberg, and the Ruhr*, 3 WASH. U. J. L. & POL'Y 849 (2000), *published in* EVOLVING VOICES IN LAND USE LAW ch. 13 (Wash. U. J. L. & Pol'y ed., 2000).

188. Julianne Kurdila & Elise Rindfleisch, *Funding Opportunities for Brownfield Development*, 34 B.C. ENVTL. AFF. L. REV. 479 (2007).

1. Local Response

Local government can provide subsidy for brownfield development by offering faster processing of development review and lowering infrastructure-based impact fees and by supporting development through loans, grants for clean up, or equity participation as a partner. Zoning and other regulations should be liberalized to encourage mixed-use, transit-linked development, offering more housing units at higher densities and incentives to reduce on-site parking while assuring adequate environmental safety. Brownfield redevelopment will be encouraged by imposing restrictions on developing open space on the urban periphery. Urban growth boundaries are essential to redirecting development to infill and urban redevelopment of brownfields.[189] Brownfields can also make excellent sites for renewable energy such as biomass or a wind farm at a landfill.[190] Recycling and waste management can eliminate the need for out-of-town disposal transport by truck or pipe. What can not be recycled can fuel zero-emissions cogeneration plants producing heat, hot water, and energy.

2. Regional Response

Planning at the regional level could establish an integrated plan for brownfields, allowing the sharing of tax base among communities.[191] Counties or regional consolidated government entities could allocate tax revenues to reward and subsidize the additional costs faced by brownfields to provide urbanization sites in municipalities or for parks or urban agriculture. Traditionally, without regional tax base sharing, the benefits of development are enjoyed by municipalities containing housing settlements for the affluent and industrial, commercial, or office projects that generate generous tax revenues. Like municipalities, counties and regional governmental entities can contribute to the funding of clean-up. Only at the regional level can transportation be planned and coordinated to establish priorities and a unified system of higher density corridors among cities and suburbs. The design should be one of polycentric

---

189. Michael Lewyn, *Sprawl, Growth Boundaries and the Rehnquist Court*, 2002 UTAH L. REV. 1; Robert Stacey, *Urban Growth Boundaries: Saying "Yes" to Strengthening Communities*, 34 CONN. L. REV. 597 (2002).

190. Steven Ferrey, *Converting Brownfield Environmental Negatives into Energy Positives*, 34 B.C. ENVTL. AFF. L. REV. 417 (2007).

191. ORFIELD, *supra* note 168 at 105–108; RUSK, *supra* note 168; Note, *Making Mixed-Income Communities Possible, supra*, note 168.

nodes tying together suburban districts, city and town centers. Regions are best able to resolve the larger issues of disposing contaminated soil, water and other materials.

3. State Response

States are in the best position to finance brownfield cleanup and encourage redevelopment or reuse of existing structures. The state can establish an agency to provide loans and grants for brownfield development and establish a statewide plan for such reuse providing adequate incentives and a fair apportionment of resources to reduce local government competition for favored uses. The state can also make difficult choices when local government consensus is not possible, e.g., placement of industries that include waste or trucking, prisons, and other disfavored uses such as housing (other than housing for the affluent). States can also cooperate with adjacent states to establish interstate efforts to make allocation decisions between states and municipalities. Redevelopment laws can be expanded to allow and subsidize tax increment-financed redevelopment and enterprise zones establishing reduced taxation for brownfield development and the support of transit-oriented development at or adjacent to brownfield redevelopment.

4. Federal Response

The Congress can increase funding available under the Comprehensive Environmental Response, Compensation, and Liability Act of 1980 (CERCLA),[192] otherwise known as the Superfund law, to provide subsidies for toxic dumps and brownfields clean up. In addition to providing funding for cleanup of pol-

---

192. 42 U.S.C. §§ 9601–9675 (2000). *See generally* Ralph A. DeMeo et al., *Insuring Against Environmental Unknowns*, 23 J. LAND USE & ENVTL. L. 61 (2007) (discussing environmental risk in brownfield redevelopment); Bradford C. Mank, *Reforming State Brownfield Programs to Comply with Title VI*, 24 HARV. ENVTL. L. REV. 115, 120 (2000). *See also* Casey Cohn, Student Article, *The Brownfields Revitalization and Environmental Restoration Act: Landmark Reform or a "Trap for the Unwary"?*, 12 N.Y.U. ENVTL. L.J. 672 (2004); Amy Pilat McMorrow, Note, *CERCLA Liability Redefined: An Analysis of the Small Business Liability Relief and Brownfields Revitalization Act and its Impact on State Voluntary Cleanup Programs*, 20 GA. ST. U. L. REV. 1087 (2004); Kashif Haque, Note, *Internal Revenue Code Section 1989, the Tax Incentive for Brownfield Redevelopment: A Sheep in Wolf's Clothing*, 8 WASH. U. J. L. & POL'Y 371 (2002).

luted sites, Congress could offer subsidies for transit-served redevelopment projects, particularly where they are designed to include affordable housing and support to small business development. Subsidies can be used to reduce commercial rents to encourage the development of complementary development, adequate shops to support residents, and new entrepreneurs. Congress is also able to award grants based on the size of brownfields, the wealth or poverty of the region, and to those communities that are aggressively pursuing brownfield policies and plans.

5. International Response

Internationally, the United States can target foreign aid for use in supporting communities to reduce sprawl, cleanup brownfields, and assist redevelopment. The formation of public, transit-served, compact cities without automobiles will dramatically reduce carbon emissions. The international community can set standards for urbanization policies and share resources to assist poorer nations in addressing brownfields to densify and fully utilize centrally-located sites.

Chapter 5

Consumption and Conservation

> The climate crisis presents us with an inconvenient truth. It means we are going to have to change the way we live our lives. Whether these changes involve something as minor as using different lightbulbs, or as major as switching from oil and coal to other fuels, they will require effort and cost money. But many of these needed changes will actually save money and make us more efficient and productive. We all must take action so that our democracy creates laws to protect our planet, because we simply can't afford not to act.
>
> AL GORE, AN INCONVENIENT TRUTH 179 (2006)

> We consume as if there's a prize for consumption itself, like the pie-eating contest at the county fair.
>
> TERRY TAMMINEN, LIVES PER GALLON: THE TRUE COST OF OUR OIL ADDICTION 208 (2006)

Although the emphasis on carbon emissions is often energy production, manufacturing, and transportation, an extraordinary component is tied to the cumulative effect of individual consumption. The reader should not assume that modifying individual consumption patterns will have little effect on the environment or the future of civilization. Although power generation, vehicles, manufacturing processes, and construction generate the greatest share of greenhouse gases, individuals can cumulatively reduce demand and affect a modification of how we build, generate power, and transport ourselves. This chapter is meant to offer a primer and overview of all the ways we can reduce emissions. The average American individual's share of total emissions in 2000 was more than 14,000 pounds of carbon dioxide, for a total of 4.1 trillion pounds for all Americans.[193] By comparison, all of American industry emitted

---

193. Michael P. Vandenbergh & Anne C. Steinemann, *The Carbon-Neutral Individual*, 82 N.Y.U. L. REV. 1673, 1677 (2007).

3.9 trillion pounds in 2000. The 4.1 trillion pounds attributable to American individuals comprise roughly 32% of total U.S. annual emissions, and 8% of the world total. It is larger than the emissions from all of Africa, Central America, and South America combined, and larger than the emissions of every foreign country besides China.[194]

The ecological footprint is a measurement of one's consumption of nature's resources.[195] As of 2000, on average, there were 5.3 acres of land for each person on the planet. The average person, however, uses 6.9 acres worth of resources and ecoservices, demonstrating overconsumption of resources.[196] Amazingly, in the United States, the average resident uses 24 acres' worth.[197] By 2040, because of population growth and rising seas, the average number of acres per person will drop to 3.5, causing the world ecological deficit to rise rapidly.[198] An ecological deficit results if the number of acres needed to sustain a nation's population is greater than the number of acres in the country.[199] The United States has a deficit of 11 acres per person.[200] As futurist, James Martin has warned:

> The Earth's ecological deficit can't last. We are using more water than the rain can renew, catching more fish than are spawned, cutting down more timber than we can regrow, pumping more carbon dioxide into the atmosphere than can be absorbed and depleting topsoil that took tens of thousands of years to accumulate. Even if people with good management skills stop doing this, billions may not. We can fight the ravages described by Malthus if we have good management, but the world's massive population growth will be in the countries least capable of managing their water, agriculture, fisheries and forestry. Ultimately, managing these sustainably is a nonnegotiable condition for life.[201]

By 2001, the biomass deficit had reached 20% and our population was at 6 billion.[202] Currently, the planet is continuing to consume 20% more biomass (all living things) than is sustainable and the population has reached 6.6 billion.[203] In economic terms, humans reached the Earth's carrying capacity, based

---

194. *Id.*
195. MARTIN, THE MEANING OF THE 21ST CENTURY, *supra* note 32 at 49.
196. *Id.*
197. *Id.*
198. *Id.*
199. *Id.*
200. *Id.*
201. *Id* at 49–50.
202. FLANNERY, *supra* note 4 at 79.
203. *Id* at 78.

on resource demand, in 1986, and since that time we have been running the environmental equivalent of a deficit budget, sustained by plundering our capital base.[204] By 2050, if humans can be found, they will require two planets' worth of resources to power the Earth.[205]

The decision of what type of vehicle one uses, the location of one's home and the type of commute, i.e., automobile, public transport, walking or biking, has a profound effect. The carbon footprint reflects the amount of carbon emissions generated by each person. The average American annually generates 9107 pounds of carbon dioxide constituting 63% of total annual emissions.[206] The materials used to construct a home, how large that home is, whether an existing structure is reused and what form of energy is used for heating and cooling influences emissions. Even choices of household members has an impact, e.g., a family of four lives more efficiently compared with those who reside alone. One refrigerator or stove, rather than four of each, can make a difference. Selecting energy-efficient appliances can also cumulatively make a difference. Federal government regulations have increased the efficiency and reduced emissions from certain appliances.[207] The typical household spends $1,900 a year on energy bills but by upgrading to energy-saving Energy Star-qualified appliances, a 30% reduction, or more than $600 per year, can be realized.[208] Energy Star-generated carbon savings in 2005 equaled the emissions from 23 million cars.[209] Replacing the refrigerator alone can make a big difference because it uses more energy than any other appliance—Energy Star-qualified refrigerators use 40% less energy than models produced before 2001.[210] Simply not opening the door of the refrigerator could save between $30 and $60 annually, saving enough electricity to light every home in America for four and one-half months.[211] Energy Star air conditioning systems use between 20 and 40% less energy. If all U.S. households switched from hot-hot to warm-cold cycles on their washing machines, the energy equivalent of 100,000 bar-

---

204. *Id* at 78–9.
205. *Id* at 79.
206. Vandenbergh & Steinemann, *supra* note 193 at 1678.
207. Dernbach, *U.S. Policy*, in GLOBAL CLIMATE CHANGE AND U.S. LAW *supra* note 185 at 61, 70.
208. *Top Tips to Stop Global Warming*, AOL.COM, Sept. 2, 2007, at 6, *available at* http://reference.aol.com/planet-earth/global-warming/top-tips-stop-global-warming.
209. ELIZABETH ROGERS & THOMAS KOSTIGEN, THE GREEN BOOK: THE EVERYDAY GUIDE TO SAVING THE PLANET ONE SIMPLE STEP AT A TIME 127 (2007).
210. *Top Tips to Stop Global Warming, supra* note 208 at 6.
211. ROGERS & KOSTIGEN, *supra* note 209 at 4–5.

rels of oil would be saved daily.[212] Drying clothes on a clothesline rather than an electric dryer on warm dry days can save energy and produce superior laundry, although the practice is illegal for the 60 million people residing in association-governed communities.[213] Not overloading an electric clothes dryer saves 5% of the electric bill, a practice that cumulatively could save 350 million gallons of gas annually.[214] A home energy audit is often free from utilities providers and reports include how much energy a home uses each year and provide tips on how to reduce consumption.[215] The average American family can cut its carbon dioxide emissions by 1,000 pounds each year.[216] Up to 20% of heating and cooling energy is lost due to poorly sealed or insulated ducts in a home.[217] Leaving a fireplace damper open can lose 8% of a home's heat and lose cool air in the summer, costing $100 annually.[218] If every American kept curtains closed on windows when the weather is warm in summer or cold in winter it could reduce energy needs by up to 25% and the total annual energy savings would equal Japan's annual energy use.[219] Tax credits on efficiency-increasing heating and cooling equipment are available, allowing a $300 credit on central air conditioning units and up to 30% on solar water heaters.[220] For each degree that Americans raise their thermostat for air conditioning and lower it for heating, $10 billion annually could be saved, enough to provide a year's energy to Iowa.[221]

What one eats can also have a significant effect, i.e., where the food comes from and, whether it is local or regional or has been transported from another continent or driven long distances by truck. Energy consumed in cooking is also relevant; the cumulative conversion to less cooked foods and foods that require minimal cooking energy can be significant. Microwaves are more than four times more energy efficient than traditional ovens.[222] If everyone in North

---

212. *Id* at 9.
213. Anne Marie Chaker, *Clothesline Has Neighbors Bent Out of Shape in Bend*, REAL ESTATE JOURNAL.COM, Sept. 24, 2007.
214. ROGERS & KOSTIGEN, *supra* note 209 at 9.
215. *Top Tips to Stop Global Warming, supra* note 208 at 1.
216. *Id.*
217. *Id* at 7.
218. ROGERS & KOSTIGEN, *supra* note 209 at 7.
219. *Id* at 8.
220. *Top Tips to Stop Global Warming, supra* note 208 at 7.
221. ROGERS & KOSTIGEN, *supra* note 209 at 2.
222. *Id* at 4 (between 3.5 and 4.8 more energy efficient); *Fire Up the Microwave*, AOL.COM. Sept. 2, 2007, at 2, *available at* http://coaches.aol.com/kids-and-family/kostigen-rogers/go-green-save-money (excerpted from ROGERS & KOSTIGEN, *supra* note 209).

America cooked exclusively with a microwave for a year, the energy savings would equal as much energy as the continent of Africa consumes during that time.[223] If everyone used the correct size pot the energy savings would equal the energy consumed by an African over the course of their life and save $36 annually for an electric range, $18 annually for a gas range.[224] Ordering and preparing less food can reduce waste and use of power and oil dramatically. On average, every American throws away about 12 pounds of uneaten poultry annually.[225] If 30% of American households could reduce oven preheating by one hour per year, the energy saved could bake a dozen cookies for every American.[226] If over the course of a year, each household purchased just one less pound of chicken, the total water saved by not having to produce and package it would be 66 billion gallons—more than all of California uses in a week.[227] Using perishable ingredients and saving leftovers could save 20 pounds of food annually and if all households followed a policy of saving food, it would save an amount sufficient to provide three meals a day for the entire year to the 1.35 million homeless children.[228] Purchasing block cheese instead of individually wrapped slices of American Cheese can save 13.8 million gallons of gasoline per year.[229]

The average person in the U.S. drinks eight ounces of bottled water per day requiring 1.5 million barrels of oil annually to produce all the plastic bottles of which could be saved by using and refilling a single water bottle.[230] Exchanging the use of plastic bottles for a water filter saves 1.5 tons of plastic used to bottle 89 billion liters of water, enough plastic to put filters in every home on the planet[231]

Each person in the U.S. produces 4.54 pounds of trash daily or 1,657 pounds per year which as a nation amounts to 500 billion pounds annually.[232] Reducing garbage by 25% will reduce carbon dioxide emissions by 1,000 pounds per

---

223. ROGERS & KOSTIGEN, *supra* note 209 at 4.
224. *Id* at 5.
225. *Go Green Save Money*, AOL.COM. Sept. 2, 2007, at 11, *available at* http://coaches.aol.com/kids-and-family/kostigen-rogers/go-green-save-money.
226. ROGERS & KOSTIGEN, *supra* note 209 at 4.
227. *Go Green Save Money, supra* note 225 at 11.
228. ROGERS & KOSTIGEN, *supra* note 209 at 3.
229. *Id* at 67–68.
230. *Id* at 32; Breanne Gilpatrick, *Cities Push Tap Water as 'Better Than Bottled,'* MIAMI HERALD.COM, Oct. 11, 2007, *available at* http://www.miamiherald.com/news/miami_dade/v-print/story/267546.html.
231. ROGERS & KOSTIGEN, *supra* note 209 at 5.
232. *Id* at 64.

year.[233] Recycling aluminum cans, glass bottles, plastic, cardboard and newspapers can reduce a home's output by 850 pounds of carbon dioxide per year.[234] If Americans recycled, separating paper, plastic, glass, and aluminum, waste sent to landfills would decrease by 75%.[235] Reusing trash bags as trash can liners can save money; a ton of plastic bags require a ton of oil to produce; a ton of paper bags require 17 trees.[236] If Americans composted their kitchen scraps instead of placing them in the trash, organic waste diverted from landfills could make a three-foot high compost pile to cover the city of San Francisco.[237] Purchasing rechargeable batteries could save the 179,000 tons of batteries discarded in the U.S. annually.[238] Sixty million plastic water bottles are discarded daily in America, suggesting the more strictly regulated tap water is safer and cheaper than the $8 billion annually spent over the cost of tap water for bottled water.[239] Tap water must meet extensive purity standards and typically is superior to bottled water.[240] A carbon footprint of zero can be accomplished by reducing carbon emissions and purchasing offsets for the remaining emissions.[241]

As the world becomes more affluent, habits change and more water is consumed. Taking a shower or using a dishwasher uses more than 9 gallons of water; taking a bath uses approximately 20 gallons, while using a bowl to wash uses just 1 gallon.[242] Each time a toilet is flushed, it uses 2.6 to 4.5 gallons of water.[243] Every two minutes saved from a shower can conserve 10 gallons of water.[244] Forty percent of drinking water supplied to homes in the U.S. is flushed down the toilet.[245] Turning off the tap while you brush your teeth can save each person 5 gallons of water daily, while not waiting for hot water to shave can save 1,825 gallons annually.[246] If each American saved one gallon of water

---

233. *Top Tips to Stop Global Warming, supra* note 208 at 9.
234. *Id.*
235. ROGERS & KOSTIGEN, *supra* note 209 at 2.
236. *Id* at 5.
237. *Id* at 3.
238. *Id* at 16.
239. *Id.*
240. Gilpatrick, *supra* note 230.
241. Vandenbergh & Steinemann, *supra* note 193 at 1677.
242. PAUL MASON, PLANET UNDER PRESSURE—POPULATION 28 (2006) (2.6 gallons); ROGERS & KOSTIGEN, *supra* note 209 at 6 (4.5 gallons what the average African uses daily for all uses).
243. MASON, *supra* note 242 at 28 (2.6 gallons); ROGERS & KOSTIGEN, *supra* note 209 at 6 (4.5 gallons what the average African uses daily for all uses).
244. ROGERS & KOSTIGEN, *supra* note 209 at 2.
245. *Id.*
246. *Id* at 6.

daily, it would save one hundred billion gallons annually.[247] In 2000, the average American used 56,796 gallons of water as compared to the average of 26,945 in France, 10,567 in the UK, 5,283 in China, and 131,029 in Australia.[248] Running full loads in the dishwasher and not pre-rinsing can save 20 gallons of water per load, or 7,300 gallons a year, which is the average a person drinks in a lifetime.[249]

Water-efficient toilets, showers, and appliances will help in the face of declining supplies of fresh water.[250] Installing low-flow plumbing can reduce water use of each household by 54,000 gallons annually, enough for a bottle of water for everyone on the planet.[251] Dual-flush toilets with options of a 0.8 gallon or a 1.6 gallon flush can reduce water use up to 67% over the traditional three gallon toilet.[252] Replacing an older toilet can save about 7,500 gallons of water a year, and fixing a leak in a toilet can save as much as 200 gallons a day.[253] Drip irrigation of yards rather than sprinklers reduces evaporation and can reduce water use by 70%.[254] A rain sensor that overrides automatic sprinklers during and following rain could drop annual water use by up to 30%.[255] An automatic shut-off nozzle for garden hoses can save 6.5 gallons per minute.[256] A pool cover can reduce evaporation by 90%.[257]

Similarly, reusable household items as compared to nonreusable disposable materials have an effect. Using reusable bags when shopping, as the Europeans do, helps towards saving the 12 million barrels of oil it takes to produce the plastic bags Americans use each year; 15 million trees are cut down to produce paper for paper bags.[258] If just 10% of U.S. households used paperboard rather than plastic spindled double-tipped cotton swabs, 150,000 gallons of gasoline would be saved annually.[259] If one in seven U.S. households purchased shampoo with conditioner, the plastic saved annually could fill a football field 27-

---

247. *Id* at 6–7.
248. MARTIN, THE MEANING OF THE 21ST CENTURY, *supra* note 32 at 29.
249. ROGERS & KOSTIGEN, *supra* note 209 at 3.
250. *Id.*
251. *Id* at 127.
252. *Id* at 130.
253. *Top Tips to Stop Global Warming*, *supra* note 208 at 10.
254. ROGERS & KOSTIGEN, *supra* note 209 at 10–11.
255. *Id* at 11.
256. *Id.*
257. *Id.*
258. *Top Tips to Stop Global Warming*, *supra* note 208 at 4.
259. ROGERS & KOSTIGEN, *supra* note 209 at 79–80.

stories high.[260] If 10% of households chose alternatives to aerosol shave gels and foams, the petroleum savings from the propellants could light 270,000 households for a month.[261]

Between 10 and 15% of the energy used by a television is still used when it's powered "off." If every home unplugged televisions when they were not being used by connecting them to a wall-switched outlet, the nation would save more than $1 billion in energy bills.[262] If 5% percent of households chose a 32-inch LCD panel over an equal size plasma screen, the energy saved could power all of the 266 million televisions for 40 straight hours.[263] Unplugging other appliances would save more than $5 billion annually for this standby power alone, about 5% of all electricity consumed in the country.[264] Installing "auto switching" power strips in every home to shut down the primary appliances when not in use would save enough energy to power 40,000 homes for a year.[265] Setting hot water at 120° F will reduce fuel consumption, which could cumulatively save $32 billion annually in energy costs.[266] The excessive burning of lights both for security and advertising not only causes light pollution, hindering star gazing, they consume energy that generates carbon and heat.[267] Views from space vividly disclose illumination around the globe. Businesses can cease unnecessary lighting and use minimal security lighting. Individual consumers can reduce excessive lighting and power-consuming holiday displays.

Sending e-mails from a computer uses 30 times the electricity for each message as compared to phone text messages, especially quick, one-line notes, from a hand-held device or cell phone.[268] Since laptop computers use 50% less energy than desktops, if every desk top computer in the U.S. was exchanged for a laptop, the savings would be $2.5 billion in energy costs.[269] If one-quarter of U.S. households replaced conventional computers with Energy Star-qualified computers that adjust to lower power mode when not in use, the energy

---

260. *Id* at 100–101.
261. *Id* at 104.
262. *Id* at 19–20.
263. *Id* at 93.
264. *Id* at 28.
265. *Id* at 37.
266. *Id* at 9–10.
267. David Hughes & Martin Morgan Taylor, *And Can't Look Up and See the Stars*, 16 J. ENVTL. L. 215 (2004); Kristen M. Ploetz, Note, *Light Pollution in the United States: An Overview of the Inadequacies of the Common Law and State and Local Regulation*, 36 NEW ENG. L. REV. 985 (2002).
268. *Go Green Save Money, supra* note 225 at 8.
269. ROGERS & KOSTIGEN, *supra* note 209 at 39, 91.

saved over those computers' lifetimes could light every home in the U.S. household for more than a year.[270] Laser printers consume 30 times the electricity of an inkjet printer.[271]

What we wear affects emissions. Imported items and transported items require more energy and some materials may call for different production methods and energy consumption. For example, organic cotton requires less energy than non-organic cotton that uses fertilizers or pesticides; alternative fabrics such as used clothing, polar fleece made from recycled plastic, and the purchase of fewer clothing items would also be more sustainable. Organic cotton does not have pesticide residue, but it and all fabrics are produced unsustainably using large amounts of water and energy, including "Eco Intelligent Polyester," partially-recycled fabrics and nylon, bamboo and hemp products.[272] Used clothing can lessen the need for water consumption and deforestation to make way for cotton fields.

Some will claim they live a carbon-neutral lifestyle by purchasing carbon offsets such as through investment in green energy, but living in a large home, driving an SUV, or traveling frequently by jet cannot be justified by carbon offsets.[273] Al Gore's "An Inconvenient Truth" documentary became carbon neutral by computing the carbon generated in production and paying under $500 to a broker to invest in green energy.[274] As yet, however, there have been no actual emissions reductions resulting from the purchase of or investment in carbon credits.[275] Being overweight or obese requires more fuel in a car, airplane, or other transport, and transporting heavy luggage consumes more fuel. Walking, bicycling, exercise, and weight loss will help forestall global climate change, as will ceasing the consumption of foreign-made goods. Reductions in unnecessary travel, particularly by automobile and airplane, will assist in reducing carbon emissions. Until air travel is eliminated for personal and business

---

270. *Id* at 90.

271. *Id* at 59.

272. *See* Umbra Fisk, *The Environmentalist's New Clothes*, GRIST, July 12, 2004, available at http://www.grist.org/advice/ask/2004/07/12/umbra-clothing/index.html.

273. Allison Linn, *Carbon Offset Market Raises Questions*, MSNBC, May 22, 2007, available at http://www.msnbc.msn.com/id/18659716/; Alan Zarembo, *Can You Buy a Greener Conscience?: A Budding Industry Sells 'Offsets' of Carbon Emissions, Investing in Environmental Projects. But There Are Doubts About Whether it Works*, L.A. TIMES, Sept. 2, 2007, available at http://www.latimes.com/news/science/environmental/la-sci-offsets2sep02,1,5021370.story?ctrack=1&cset=true.

274. Zarembo, *supra* note 273.

275. *Id.*

travel or freight transport, limiting passenger travel to necessary trips, eliminating vacation packages involving air flight, and strictly limiting baggage weight will help reduce emissions. Increasing the taxation of air travel, eliminating budget flights, and charging passengers by weight will also more fairly apportion costs and restrict excessive emissions. A range of organizations offer "carbon offsets." These are donations of funds used to invest in renewable energy projects and planting trees to offset emissions caused by the activity that motivated the donation.[276] Making up for emissions probably isn't as expensive as one would think. The emissions from two air flights can be offset for under $10 per person.[277] A better option is to eliminate activities with significant carbon emissions, however.

Other actions consumers can take include eliminating aerosol sprays in favor of pump spray dispensers, reducing trash and waste, and living a more minimalist rather than a materialist lifestyle with possessions limited to necessary and long-lasting items. For example, a .33-ounce eighteen-karat-gold band requires more than 13,000 gallons of water to produce and leaves behind twenty tons of cyanide-laden mine sludge. If one in one thousand households purchased antique, recycled, or vintage jewelry for their next purchase the savings would be two million tons of mine waste and 1.37 billion gallons of water.[278] Since the average U.S. household receives 1.5 trees' worth of junk mail each year, registering with "opt out," a free call service, or with Mail Preference Service for $1 can save a lot of trees from being thrown right into the trash.[279] Prepackaged foods are likely to score high in carbon emissions and should be avoided due to the excessive packaging. When picking up take out food, consumers should request only what is needed in packaging, napkins, plastic utensils or chopsticks, and batches of condiments, since millions end up in the trash.[280] If every household refused a paper ATM receipt and ordered paperless statements, the savings would allow more than 17,000 high school graduates to attend public university for a year.[281] If every American household paid bills online, it would reduce solid waste by 1.6 billion pounds and cut greenhouse gas emissions by 2.1 million tons each year.[282] If everyone el-

---

276. *Top Tips to Stop Global Warming, supra* note 208 at 11.
277. *Id.*
278. ROGERS & KOSTIGEN, *supra* note 209 at 89.
279. *Id* at 7.
280. *Go Green Save Money, supra* note 225 at 10.
281. ROGERS & KOSTIGEN, *supra* note 209 at 119–20 (ATM receipts are the leading source of litter on the earth potentially forming a roll of paper that would circle the equator 15 times).
282. *Top Tips to Stop Global Warming, supra* note 208 at 5.

igible made automatic bank deposits of paychecks, $65 billion would be saved in fuel costs and lost time and the paper saved would be enough to supply paper checks for every person in the world.[283] Telephone books comprise almost 10% of all waste in dump sites and can easily be replaced by the use of on-line directories.[284] Burning of all kind generates carbon emissions whether it is a fireplace, a campfire, a barbeque, or smoking of all kinds. If, however all smokers would drop the 1.5 billion butane lighters made of petroleum that annually end in landfills or incinerators and switch to cardboard book matches that are made of recycled paper over wood matches from trees, 5.5 million trees could be saved annually.[285]

If you must own a car, think hybrid, but a used car will save even more energy—as well as over 2,150 pounds of steel.[286] If one in a one hundred potential new car buyers chose a used car instead, the amount of steel saved annually could reconstruct the Golden Gate Bridge, twice a year.[287] Switching to a car that gets 10 more miles to the gallon can save more than $1,000 dollars a year; hybrid vehicles like the Toyota Prius get an estimated 60 highway miles to the gallon.[288] Trading your vehicle for a hybrid will reduce your personal carbon emissions by 70%.[289] Driving 55 mph can save you more than 20% on the consumption and the cost of gas.[290]

If just 10% of the 65 million Americans that garage their car, and thus do not require lengthy engine warming, idled their car for five fewer minutes each day, the savings would be 84.5 million gallons of gas annually, enough for a million people to drive across the country.[291] Washing a car in a commercial car wash uses up to 100 gallons less water than washing it at home and the commercial car washes often recycle and reuse rinse water.[292] If all drivers made the switch, some 12 billion gallons of soapy water could be diverted from entering the country's waterways.[293] Properly inflating tires can save 2–3% of fuel; although best gas milage requires windows to be closed, opening windows to eliminate hot air when a vehicle is hot and running the air conditioning on

---

283. ROGERS & KOSTIGEN, *supra* note 209 at 119.
284. *Id* at 9.
285. *Id* at 8.
286. *Go Green Save Money, supra* note 225 at 12.
287. ROGERS & KOSTIGEN, *supra* note 209 at 87.
288. *Top Tips to Stop Global Warming, supra* note 208 at 2.
289. FLANNERY, *supra* note 4 at 305.
290. *Top Tips to Stop Global Warming, supra* note 208 at 2.
291. ROGERS & KOSTIGEN, *supra* note 209 at 10.
292. *Id.*
293. *Id.*

lower settings can save another 2–3%.²⁹⁴ A few more percent can be saved by eliminating jackrabbit starts; 1–2% can be saved by using "Energy Conserving II" motor oil.²⁹⁵ If just 5% of households used rerefined motor oil that produces five quarts from two gallons of used oil rather than virgin oil for oil changes that requires two barrels of crude oil to produce the same five quarts, 2.5 billion gallons of oil would be saved annually.²⁹⁶ Maintaining an empty trunk space can increase milage per gallon of gas.²⁹⁷ Topping off the tank causes expanding warming gas to discharge through vents; idling wastes gas.²⁹⁸ Loading a roof rack can cut mileage by 5%; each 100 pounds of cargo reduces fuel economy by 1–2%.²⁹⁹ High octane gas wastes extra energy required to produce the fuel and should not be used unless the owner's manual requires its use.³⁰⁰ Using retread tires saves 60 gallons of oil.³⁰¹ If 10% of the nation bought retreads, the 290 million gallons saved annually would save the $900 million worth of gasoline used daily by all the cars in the U.S.³⁰² Telecommuting or using public transport even one day per month can save 5% of fuel.³⁰³ If Americans would use public transit at the rate of Europeans for 10% of daily travel, the energy saved would equal all the energy used annually by the U.S. petrochemical industry and nearly all the energy used to produce food in the nation.³⁰⁴ If households could reduce the use of lawn fertilizers by 25% or three pounds annually by leaving grass clippings on the lawn after mowing, it would save 1.3 billion pounds of chemical fertilizers and more diesel fuel than Amtrak uses in six years.³⁰⁵ Gas-powered garden tools such as mowers emit 5% of the nation's air pollution, using 6 hundred million gallons of gasoline each year.³⁰⁶ One gas-powered lawn mower emits 11 times more air pollution than the average new car each hour of operation.³⁰⁷ Trading up to a hybrid, utilizing green

---

294. TAMMINEN, *supra* note 148 at 162–63.
295. *Id.*
296. ROGERS & KOSTIGEN, *supra* note 209 at 86.
297. Chris Backes & Reinske Teuben, *Legal Aspects of the Dutch Approach to CO_2 in* CLIMATE CHANGE AND THE KYOTO PROTOCOL: THE ROLE OF INSTITUTIONS AND INSTRUMENTS TO CONTROL GLOBAL CHANGE 128, 129 (Michael Faure *et al.* eds. 2003).
298. TAMMINEN, *supra* note 148 at 162–63.
299. *Id.*
300. *Id.*
301. ROGERS & KOSTIGEN, *supra* note 209 at 87.
302. *Id.*
303. TAMMINEN, *supra* note 148 at 163.
304. *Id.*
305. ROGERS & KOSTIGEN, *supra* note 209 at 76.
306. *Id* at 133–34.
307. Id.

power options offered by power companies and voting for politicians committed to carbon emission reduction can change the world, although it might not avoid extinction.[308]

1. Local Response

Local ordinances could tax or restrict the sale or importation of imported goods or those that involve significant transportation. Similarly, disposable goods and packaging could be restricted and taxed while aerosols and inefficient light bulbs can be prohibited. Communities should prohibit the sale of bottled water which requires fleets of trucks to haul and distribute; contributes to traffic congestion, air pollution, carbon emissions; and the use of billions of plastic bottles.[309] Ordinances can assure replacement of old incandescent light bulbs with compact fluorescent light bulbs (CFLs) that will help increase energy efficiency and lower lighting costs for replacing one old incandescent bulb with a CFL can save up to $30 over the life of the bulb (generally five years) and CFLs use two thirds less energy, give off 70% less heat and last up to 10 times as long as conventional bulbs.[310] The sale of incandescent bulbs can be prohibited or CFLs subsidized to set lower prices. If each household would exchange five bulbs, a trillion pounds of greenhouse gases would be removed from the air, equal to eight million cars and would produce a $6 billion energy savings.[311] Global warming can thereby be partially mitigated by reducing air pollution caused by power plants. CFLs, however, contain small amounts of mercury and thus pose a hazard in disposal and can be dangerous if broken and must be carefully disposed of.[312]

Local laws can limit house sizes and impose green architecture and low energy standards and could encourage energy saving behavior through education. To reduce the excessive use of electricity, ordinances should be enacted to limit or curtail the use of lighting on automobile sales lots, commercial, industrial, and office buildings for advertising purposes and should prohibit or impose a serious limit on holiday lighting. Cities can administer programs to

---

308. FLANNERY, *supra* note 4 at 6.
309. BROWN, *supra* note 36 at 242.
310. *Top Tips to Stop Global Warming*, *supra* note 208 at 3.
311. ROGERS & KOSTIGEN, *supra* note 209 at 7–8.
312. Alex Johnson, *Shining a Light on Hazards of Fluorescent Bulbs: Energy-Efficient Coils Booming, But Disposal of Mercury Poses Problems*, MSNBC.COM Mar. 19, 2008, *available at* http://www.msnbc.msn.com/id/23694819/.

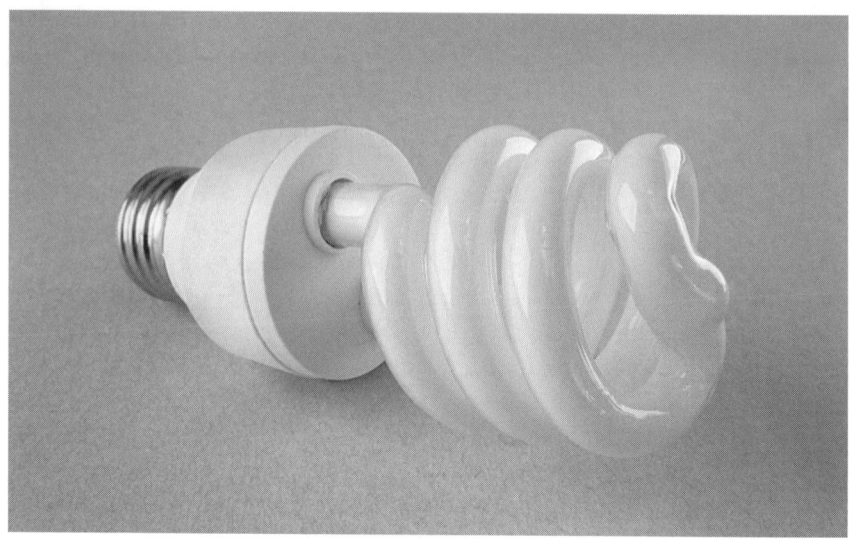

Energy Star compact fluorescent light bulb (CFL). Source: EPA/DOE.

educate consumers on conservation and offer energy surveys and discounts on energy-efficient goods and services.[313] Although dealers and manufacturers are willing to meet efficiency standards in producing and marketing appliances, only when utility companies or local government offered a $50 discount were consumers interested in the more efficient alternatives.[314]

Berkeley, California passed Measure G calling for reducing greenhouse gases from the city by 80% by 2050.[315] The city is proposing that landlords provide each tenant with a free transit pass while allowing a $7 monthly rent increase; parking for car sharing cars[316] be provided for on every block; all new buildings, resold homes, and renovations include replacing older appliances with solar-

---

313. HARRIET BULKELEY & MICHELE M. BEWTSILL, CITIES AND CLIMATE CHANGE: URBAN SUSTAINABILITY AND GLOBAL ENVIRONMENTAL GOVERNANCE 114 (2003) (describing Leicester, UK).

314. DAVID B. GOLDSTEIN, SAVING ENERGY GROWING JOBS: HOW ENVIRONMENTAL PROTECTION PROMOTES ECONOMIC GROWTH, PROFITABILITY, INNOVATION, AND COMPETITION 155–56 (2007).

315. Carolyn Jones, *It Won't Be Easy Being Green: Berkeley Sets Tough Course for its Residents to Follow to Help Reduce Emissions of Greenhouse Gases in City,* SAN FRANCISCO CHRONICLE, May 24, 2007, at A1.

316. BEATLEY, GREEN URBANISM, *supra* note 170 at 150–56; KUSHNER, THE POST-AUTOMOBILE CITY, *supra* note 170 at 99–102 ; William P. Macht, *The Rise of Car Sharing,* 62 URB. LAND 26 (Jan. 2003); Anne Marie Mannion, *In it for the Short Haul: Car-Sharing for*

powered water heaters and other high efficiency appliances, upgrading insulation and windows, providing garages with outlets for electric cars; and green requirements for all buildings including the use of recycled materials.[317] The city will establish assessment districts to fund residential solar panels.[318]

A serious problem with congestion is the large percentage of traffic that is cruising to search for a cheap, on-street parking meter or free parking. Although raising parking meter fees will increase available spaces and reduce cruising, as long as parking is more than $1.20 per hour, the cost of operating a car, it will be cheaper to cruise than park.[319]

Cherry Hill, New Jersey has joined a number of cities that offer points that can be redeemed at local businesses for recycling; after one week, recycling had doubled from 12 pounds per household to nearly 26 pounds, saving thousands of dollars in landfill fees.[320]

2. Regional Response

On a regional level, efforts should be coordinated to use locally produced building materials, household, wearing apparel, and food.

3. State Response

Congestion pricing on roads could include higher tolls on trucks transporting building materials, household goods, wearing apparel, and food beyond a set distance. Reductions for recycled materials not locally available can be established. However, recycling regionally should be the norm instead of sending paper to China for recycling or other recycling products to other distant destinations. States are in a position to offer incentives for energy and environmental conservation improvements, the development of low energy food preparation and ecologically friendly, modest homes through regulation, taxing and spending initiatives. Experience with state and federal tax credits for

---

Urban Errands Brings it to Battle Against Price, Parking and Pollution to Chicago, CHI. TRIB., Sept. 12, 2002 at N1.
    317. Jones, *supra* note 315.
    318. *Id.*
    319. SHOUP, *supra* note 171 at 275–79, 289–91, 295–315.
    320. *New Jersey Town Doubles Recycling Rates in One Week with the RecycleBank Program*, GREEN PROGRESS, Dec. 3, 2007, *available at* http://www.enn.com/pollution/article/26368.

installing solar heating, electricity generation, or insulation have resulted in exploitation by installers with no significant environmental benefits and higher priced insulation.[321] Price ranges should be established and controlled and a list of eligible specific products and installers identified.

4. Federal Response

The federal government is best able to impose tariffs or embargoes on goods that require substantial energy consumption, inefficient energy use to manufacture or transport, or that are not produced using reusable energy. Federal tax policy can also be directed to rewarding conservation and investment in low emissions manufacturing, transportation, building materials and methods. Gas guzzler taxes should be imposed for low mileage per gallon vehicles and higher taxes levied on homes that exceed a maximum square footage or fail to utilize best energy consumption policies. A program of eco-labeling whereby products are rated according to the use of environmentally sound, energy-efficient methods would enable consumers and the market to reward sustainable production.[322] Congress can build on its accomplishments in enacting The National Appliance Energy Conservation Act of 1987,[323] in setting conservation standards for consumer products and appliances and the Environmental Protection Agency's Energy Star program that sets energy efficiency criteria for appliances and electrical equipment such as computers at 10–20% higher efficiency than minimum requirements.[324] Foreign aid could also be directed to reducing energy and transport costs in manufacturing and food production in developing nations.

5. International Response

International treaties should prohibit food importation unless it is transported using an energy efficient transport mode, the exception being for hu-

---

321. GOLDSTEIN, *supra* note 314 at 254–57.
322. BROWN, *supra* note 36 at 235–38.
323. PUB. L. NO. 100-12, 101 STAT. 103, *codified at* 42 U.S.C. §§ 6291–97, 6299, 6302, 6303, 6305–6, 6308–9 (2000).
324. U.S. ENVIRONMENTAL PROTECTION AGENCY, ENERGY STAR, PRODUCT SPECIFICATIONS, ELIGIBILITY CRITERIA & PARTNER COMMITMENTS, *available at* http://www.energystar.gov/index.cfm?c=product_specs.pt_product_specs (last visited Mar. 24, 2006).

Bluewater Mall in Kent, the largest shopping mall in the UK.
Source: James A. Kushner.

manitarian purposes, and aid to provide local food and goods, low energy production methods, and conservation of resources. An Earth restoration budget could be established, as Lester Brown advocates, to restore the Earth's productive health that would include the costs of protecting and restoring soils, forests, rangelands, oceanic fisheries, biological diversity, and halting advancing deserts.[325]

---

325. BROWN, *supra* note 36 at xi, 157–62.

Chapter 6

Economic Development

At a time when U.S. auto manufacturers spend more money on health care for their workers than steel for their cars, it's increasingly hard to make the case that cheap electricity is a major factor in keeping jobs from being exported to Asia. By contrast, a full-blown push for clean energy could unleash a jobs bonanza that would make what happened in Silicon Valley in the 1990s look like a bake sale.
Jeff Goodell, Big Coal xxi (2006)

Effective action to prevent global warming will arguably "require resource transfers greater than all of the foreign aid, multilateral and bilateral aid in current programs." Positive inducements include the transfer of technology and financial aid from the developed world to the developing world.... But ... efforts have been feeble so far, failing to provide positive inducements to strong action. Incentives will have to be strengthened mightily and given vastly more financial backing. Coercion might come in the form of trade penalties....
Paul G. Harris, Collective Action on Climate Change:
The Logic of Regime Failure, 47 Nat. Resources J. 195, 213 (2007)
(quoting T.C. Schelling, Economic Responses to Global Warming:
Prospects for Cooperative Approaches, in Global Warming:
Economic Policy Responses 197, 199 (Rudiger
Dornbusch & James M. Poterba eds. 1991)

Environmental degradation is ... undermining development and threatens future development progress.
United Nations Environment Programme, GEO4—Global
Environment Outlook ch. 1 at 4 (2007) (Main Message)

Economic development relates to policies to generate economic activity, expand employment opportunities, and generate tax revenues beyond service or infrastructure costs. The economic base shifts over time. Although it was shocking at the end of the horse and buggy economy to shift to the automobile, the economy performed and economic development flourished. Much of it surrounded the motor vehicle, truck, and public transit industry. These changes generated further economic development throughout the economy as motor vehicles dominated travel and transport. The shift from carbon generation to carbon reduction will also pose complex changes, including the likely end to the personal automobile—at least outside of rural areas, the end of most air flights, and the end of reliance on truck transport. Just as most of the United States recovered from the reduction in manufacturing (most dramatically in the rust belt) and adjusted to the transition to a service industry and a shift to importing, so it is possible to look at global warming cures as opportunities.

The conversion to the post-fossil fuel era will grow jobs, reduce costs, positively improve the economy and clean the air.[326] Estimates indicate that rather than causing injury to the American economy, compliance with Kyoto could generate $2.3 trillion in direct benefits and substantially more from indirect benefits.[327] While official statements suggest huge costs to comply with Kyoto, e.g., as the current U.S. Department of Energy estimate of $378 billion annually, the Clinton Administration estimated it at $1 billion.[328] Other studies suggest American energy bills actually would decline $530 per household annually.[329] The U.S. government and industry have largely ignored results of economic studies that report carbon emission reduction being profitable rather than causing Kyoto compliant losses.[330] A study commissioned by DTE Energy, the parent company of Detroit Edison, released in November, 2007 reports that the U.S. could cut greenhouse gases by 28% and the reductions would more than pay for themselves in lower energy bills.[331] The report points out that consumers generally ignore energy efficiency when choosing products, that land-

---

326. GOLDSTEIN, *supra* note 314. *See also* Donald M. Goldberg & Angela Delfino, *The Impact of the Kyoto Protocol on U.S. Business, in* GLOBAL CLIMATE CHANGE AND U.S. LAW 101 (Michael B. Gerrard ed. 2007).

327. GOLDSTEIN, *supra* note 314 at 235 (reflecting $2.7 trillion in energy investments and $5 trillion in energy savings).

328. FLANNERY, *supra* note 4 at 232–33. *See also* Daniel A. Farber, *Adapting To Climate Change: Who Should Pay*, 23 J. LAND USE & ENVTL. L. 1 (2007) (arguing that "emitters pay" is the most attractive of the potential allocation principles, while climate change winners pay" is the least compelling and that different types of adaptation, including "beneficiaries pay" and "public pays," at various times and places may be suited for different mixes of allocation principles).

329. FLANNERY, *supra* note 4 at 232–33.

330. GOLDSTEIN, *supra* note 314 at 236–40.

331. Matthew L. Wald, *Study How U.S. Could Cut 28% of Greenhouse Gases*, N.Y. TIMES, Nov. 30, 2007.

lords and builders are also unconcerned with energy costs and fail to spend a bit more to dramatically increase efficiency and recommends regulation and tax incentives.[332] The report recommendations emphasize conservation, finding that renewable energy supplies are relatively modest and capturing carbon dioxide from coal power plants costly and of limited effect.[333]

In addition to generating energy, the alternative renewable energy sector will likely generate an extraordinary growth in jobs. Technology development will expand with the search for alternative transport, food production and living environments. The nation can continue to do what it has done so well in the past, i.e., provide the intelligence infrastructure for the revolutions ahead for energy, transport, and urban design. The redevelopment of efficient and comfortable intercity trains and convenient public transport can generate an army of jobs. As construction and reuse in urban development fully integrate materials, systems, and designs for energy production, water use reduction and efficiency in appliances, jobs should be plentiful. Already here is a burgeoning number of green collar jobs involving technicians to install green roofs, landscaping, tree pruning, ecological restoration, green construction, and hazardous waste cleanup.[334]

The shift to a sustainable planet can generate great economic development opportunities, probably greater than the introduction of coal, rails, steam engines or automobiles. For example, entirely new industries are possible in solar, wind, wave, battery cell, and geothermal, water and irrigation efficiency; energy and water-efficient appliances; public transport; tree planting, landscaping and forestation; waste treatment including waterless, odorless composting; and the building and retrofitting of buildings for environmental performance.[335] Automobile production lines in the United States have potential for being converted to producing wind turbines.[336]

Fossil fuels and other natural resource extraction industries although generating extraordinary wealth for the extractors has caused environmental destruction and have generated poverty for the nations involved in extraction. Currently, a resource curse exists: reflecting that the result of resource extrac-

---

332. *Id.*
333. *Id.*
334. Marisol Bello, *Cities Cultivate 2 Types of Green*, USA TODAY, Dec. 12, 2007, *available at* http://www.usatoday.com/news/nation/environment/2007-12-12-green-jobs_N.htm.
335. BROWN, *supra* note 36 at 243, 248. *See also* Sue Doyle, *L.A.'s Trash Goal: No Waste by 2030*, DAILY BREEZE.COM, Jan. 21, 2008 (current technology allows all trash to be recycled, composted, or used as alternative energy fuel in Los Angeles, with conversion, however, on a slow track).
336. BROWN, *supra* note 36 at 260.

Tram in Budapest, Hungary.
Source: James A. Kushner.

tion worldwide is poverty.[337] The beneficiaries of oil, gas, coal, and other mineral extraction have largely been the owners and investors.[338]

The high cost of atmosphere cleaning and planet saving is best viewed as an intelligent and unavoidable investment opportunity even though not all expenditures and changes may be beneficial. Increasingly economists and other experts will be needed to analyze alternatives, test and evaluate choices, and recommend the most efficient strategies and solutions.

1. Local Response

Local municipalities possess the ability to redesign the urban tapestry to address ecological and climate demands, e.g., redesigning streets around pedestrians and public transport; replacing building and zoning codes to reflect

---

337. GOODELL, BIG COAL, *supra* note 8 at 28–36.
338. *Id* at 29–36.

practices in ecological development design; planning for new high-speed and local trains; and increasing the land supply for gardens, parks, tree planting and urban agriculture will generate a new array of employment classifications. Planning commissions have a large role in designing the sustainable plan. Academia has not offered much assistance on designing sustainable plans and instead is too often defending 20th Century planning or merely offering criticism of most things including reform strategies. Local government also has the ability to establish standards for the city's infrastructure, development, and construction, and impose those standards on contractors doing business with the city. Cities are best able to organize recycling programs and engage in community education on consumption and ecological living.[339]

Although local government is not typically involved in foreign aid, several local considerations exist. For example, local communities can raise awareness of the plight of poor nations and those most affected by global warming through community education. Fund raising at the local level targeted towards assisting particular countries or communities with carbon emission challenges would assist in educating the United States' population of the importance of conservation and international cooperation. Sister cities ventures should be undertaken with poorer communities suffering from global warming impacts. Cities are able to tap tax revenues and charitable contributions to provide assistance to poor foreign communities, such as sister cities, for conservation.

2. Regional Response

Regional planning around transport is essential to accommodate compact developments near transport and rail stops and to pool the regional tax base to allow the least exclusionary design for assuring adequate residential commercial and economic development needs.[340] Intra-regional tax sharing can replace local government competition that tends to generate communities that exclude non-tax-generating uses and result in dramatically disparate wealth and tax bases between towns. Regional planning for industrial sites and office needs served by transport is also critical. At the regional level, leadership could direct colleges and universities to engage in programs to transfer technology and education to poor nations and communities.

---

339. BILL MCKIBBEN, HOPE, HUMAN AND WILD: TRUE STORIES OF LIVING LIGHTLY ON THE EARTH ch. 2 (1995) (Curitiba, Brazil).

340. ORFIELD, *supra* note 168 at 105–108; RUSK, *supra* note 168; Note, *Making Mixed-Income Communities Possible*, *supra*, note 168.

3. State Response

Most states have the authority and access to resources to provide leadership and target aid to modify the state's education infrastructure. Thus, the states can render the educational infrastructure more efficient and better adapted to training the scientists and technicians needed for the 21st Century economy. States can enable or preempt local policies on planning, transport, and building and are in the best position to establish planning policies and mandatory components such as public transport and trains. The state can also fund energy retrofitting and influence individuals and business to pursue conservation and emissions reductions. States can finance industrial expansion through tax-free lending and other incentives such as tax credits.[341] States are also able to tap tax revenues and charitable contributions to provide assistance to poor foreign communities for conservation.

4. Federal Response

Presidential leadership, judicial deference to the emergency circumstances, and Congressional support is necessary. Congress shapes education through grants for program expansion, teacher training, and financial support. Congress is also able to demand aggressive reduction of emissions from all industries and can invest in research and development leading to sustainable planning of economic development, food policies and transport. Military investment should be limited to supporting a defensive force, including adequate support for international defense programs under the United Nations, NATO, and other international coalitions. Those in the military trained in fields essential to conversion should be recruited to teach, train, or engage in essential industries.

Emission permits can be established to allow those who generate carbon to purchase permits allowing emissions in exchange for the sellers' reduced emissions.[342] Trading systems, however, are criticized as they slow the transition

---

341. Peter D. Enrich, *Business Tax Incentives: A Status Report*, 34 URB. LAW. 415 (2002); Timothy H. Gillis, *Sixth Circuit Bans Ohio Tax Credit Under the Commerce Clause, Casting a Pall on Incentives*, 101 J. TAX'N 359 (2004); Bradford C. Spencer, Note, *Evaluating Kentucky's Investment Tax Credits in Light of* Cuno v. DaimlerChrysler, Inc., 94 KY. L.J. 161 (2005–2006).

342. GORE, *supra* note 4 at 252 (endorsing a carbon market patterned after the earlier sulphur dioxide anti-acid rain market). *See also* Jae Edmonds et al., International Emissions Trading, in CLIMATE CHANGE: SCIENCE, STRATEGIES, & SOLUTIONS 245–66 (Eileen Claussen, Vicki Arroyo Cochran & Debra P. Davis eds. 2001); Lawrence H. Goulder & Brian

from fossil fuels.[343] A program of advertising to consumers to "buy green" would likely generate a different marketing by corporations and the faster adoption of green investments and processes particularly if tax incentives were included.[344]

Foreign aid at the federal level is essential and should be targeted at programs to develop local agriculture and food production, the reduction of sprawl, alternative energy, brownfield cleanup, assistance for redevelopment, sustainable planning, and building construction. Foreign aid could also be directed to reducing energy and transport costs in manufacturing and food production and should include aid for education, particularly involving essential technology transfer and the universities necessary to provide the specialists needed for development. Foreign aid and investment should most importantly be targeted to renewable energy production and technology. Additional funding for foreign sustainable development might come from a currency transactions tax,[345] an arms trade tax,[346] environmental taxes on all polluting sources,[347] carbon taxes, taxation of internet services,[348] or a lottery system.[349]

M. Nadreau, *International Approaches to Reducing Greenhouse Gas Emissions, in* CLIMATE CHANGE POLICY: A SURVEY 115, 120–31 (Stephen H. Schneider, Armin Rosencranz & John O. Niles eds. 2002); Karen MacDonald & Zen Makuch, *Emissions Trading and the Aarhus Convention: A Proportionate Symbiosis?, in* EU CLIMATE CHANGE POLICY: THE CHALLENGE OF NEW REGULATORY INITIATIVES 125–48 (Marjan Peeters & Kurt Deketelaere eds. 2006); Marjan Peeters, *Enforcement of the EU Greenhouse Gas Emissions Trading Scheme, in* EU CLIMATE CHANGE POLICY: THE CHALLENGE OF NEW REGULATORY INITIATIVES 169–86 (Marjan Peeters & Kurt Deketelaere eds. 2006); Pring, *in* EU CLIMATE CHANGE POLICY, *supra* note 116 at 188–201; Donehower, *supra* note 116 (arguing that without the incorporation of China and the United States, the world's two largest polluters, the carbon markets may serve as a successful market tool and example of the efficiency of an open market to cost-efficiently solve environmental problems, but will do nothing to curb GHG emissions and limit the effects of climate change).

343. Goulder & Nadreau, *in* CLIMATE CHANGE POLICY, *supra* note 342 at 129–31.

344. Thomas G. Burns, *Global Climate Change: A Business Perspective, in* CLIMATE CHANGE POLICY: A SURVEY 2, 279 (Stephen H. Schneider, Armin Rosencranz & John O. Niles eds. 2002).

345. CHRISTIAN SCHABBEL, THE VALUE CHAIN OF FOREIGN AID: DEVELOPMENT, POVERTY REDUCTION, AND REGIONAL CONDITIONS 42–49 (2007). *See also* GEORGE SOROS ON GLOBALIZATION 70–72 (2002).

346. SCHABBEL, *supra* note 345 at 50.

347. *Id* at 51–54.

348. *Id* at 54.

349. *Id* at 56–58.

5. International Response

Next to reducing the excessive carbon generation of the United States, federal policy must be directed to reducing the remaining 75% of carbon generated by other nations. In 2000, American foreign aid appropriations were $9.4 billion.[350] Although the developed nations together spend $53.7 billion annually on foreign aid, they spend $360 billion annually on subsidizing their agriculture, a practice that injures the local agricultural industries of developing nations.[351] Foreign aid may be the most effective method available to assist in that essential reduction.[352] Although controversial, critics point to excessive corruption in receiving nations preventing aid from reaching its target.[353] Ubiquitously, throughout the developing and developed planet the lack of informed government leadership is the primary impediment to avoiding extinction.[354] Perhaps aid should involve direct aid, construction, and distribution rather than offering credits or simply financial aid. Programs that deliver and pour concrete to be spread by families with dirt floor homes and subsidies to families that keep their children in school and get regular health checkups have met with success in Mexico.[355]

In addition to the United States targeting foreign aid towards investment in education and technology, agriculture, and in projects to replace power-based and transport-based emissions with alternative energy development, the United States can provide leadership for international cooperation in sharing resources and technology with poorer nations and with those nations that would suffer the most from conversion. The wealthy and developed nations, those that have principally been responsible for excessive carbon emissions, must be prepared to assist the developing nations in facing the post-fossil fuel age. "Until the

---

350. CAROL LANCASTER, TRANSFORMING FOREIGN AID: UNITED STATES ASSISTANCE IN THE 21ST CENTURY 14 (2000).

351. SOROS, *supra* note 345 at 33.

352. *See generally* LANCASTER, *supra* note 350; THEODORE H. MORAN, HARNESSING FOREIGN DIRECT INVESTMENT FOR DEVELOPMENT: POLICIES FOR DEVELOPED AND DEVELOPING COUNTRIES (2006); SCHABBEL, *supra* note 345; SOROS, *supra* note 345 at 70–72; JOSEPH E. STIGLITZ & ANDREW CHARLTON, FAIR TRADE FOR ALL: HOW TRADE CAN PROMOTE DEVELOPMENT (2005); Alan S. Miller, *International Trade and Development*, in GLOBAL CLIMATE CHANGE AND U.S. LAW 277 (Michael B. Gerrard ed. 2007).

353. Steve Bonta, *The United States Should Not Increase Foreign Aid to Developing Nations*, in DEVELOPING NATIONS 110 (Berna Miller & James D. Torr eds. 2003).

354. *Id.*

355. Jocelyn Kaiser, *Money—With Strings—to Fight Poverty*, 319 SCIENCE 754 (Feb. 2008).

economic inequality between the wealthy and poor countries is addressed with some degree of sincerity, no developing country will adhere to any agreement."[356] Internationally, resources must be allocated to those nations unable to invest in renewable energy and conservation as a carrot to accompany the stick of mandatory emissions reduction.

---

356. Gelbspan, The Heat is On, *supra* note 11 at 112.

Chapter 7

Education

> If President Bush made energy independence his moon shot, in one fell swoop he would dry up revenue for terrorism, force Iran, Russia, Venezuela, and Saudi Arabia onto the path of reform—which they will never do with $50-a-barrel oil—strengthen the dollar, and improve his own standing in Europe by doing something huge to reduce global warming. He would also create a real magnet to inspire young people to contribute to both the war on terrorism and America's future by again becoming scientists, engineers, and mathematicians.
>
> Thomas L. Friedman, The World Is Flat: A Brief History of the Twenty-First Century 283–84 (2005)

Education in America has required an overhaul long before global warming or climate change was discussed. Urban public schools that fail to successfully educate most poor and minority children or offer much in the way of apprenticeship training are well documented.[357] Our universities are preparing students for jobs that either do not exist or will not be very critical to the needs of society facing us in the 21st Century.[358] Although the U.S. educational system produces an excess of business administrators, it produces so few engineers and scientists that the economy depends on foreign nationals and immigrants to teach and even take the seats in our math, science, and engineering schools.[359] The cost of education also excludes many who could make significant contributions and widens the gulf between rich and poor.[360] Public-pri-

---

357. Kunstler, *supra* note 99 at 271–74 (referring to high schools as day care for virtual adults).

358. Thomas L. Friedman, The World is Flat: A Brief History of the Twenty-First Century 261–65, 288–90, 303–05 (2005).

359. *Id.*

360. *Compare* Zelman v. Simmons-Harris, 536 U.S. 639 (2002) (describing extraordinarily poor performing Cleveland schools) *and* San Antonio Indep. School Dist. v. Rodriguez, 411 U.S. 1 (1973) (sustaining property-tax financed school system resulting in significantly higher per pupil expenditures in suburbs as compared to cities) *with* Sheff v.

vate partnerships between universities and corporations engaging in research and development should assist in training in the classroom and in the field. Such partnerships are also necessary to establish high school and community college apprenticeships that are relevant to tomorrow's jobs and needs.[361] Each of the most livable cities enjoys a strong educational infrastructure from secondary schools to highly respected colleges and universities. Since fertility rates are inversely related to the level of education, universal education would greatly reduce birth rates.[362]

A major component of a successful energy, mobility, and building conversion is launching programs to expand educational opportunities and broaden the capacity of technical schools and universities to produce the needed professionals for health care, the sciences, engineering, mathematics, artificial intelligence, environmental sciences as well as construction trades necessary to move to the post-fossil fuel, reduced carbon emissions era.[363] New technologies will be replacing existing systems in transport, housing, appliances, and industrial production requiring widespread education of those who install and maintain, as well as those who design.

O'Neill, 678 A.2d 1267 (Conn. 1996) (recognizing affirmative obligation to equalize educational opportunities between city and suburb); James K. Gooch, *Fenced In: Why* Sheff v. O'Neill *Can't Save Connecticut's Inner City Schools*, 22 QLR 395, 397 (2004); Mildred Wigfall Robinson, *Fulfilling Brown's Legacy: Bearing the Costs of Realizing Equality*, 44 WASHBURN L.J. 1 (2004); James E. Ryan, *Schools, Race, and Money*, 109 YALE L.J. 249 (1999) (ineffective financial reforms); James E. Ryan & Michael Heise, *The Political Economy of School Choice*, 111 YALE L.J. 2043 (2002). *Cf.* Goodwin Liu, *Education, Equality, and National Citizenship*, 116 YALE L.J. 330 (2006) (arguing that the citizenship clause of the fourteenth amendment requires equality in educational resources between states which is a greater problem that disparity of resources between districts within states and an affirmative obligation of congress to fund a minimum floor).

361. GRAHAM HALLETT, THE SOCIAL ECONOMY OF WEST GERMANY ch. 8 (1973) (describing practical education and apprenticeship programs as part of general education program description); ARTHUR HEARNDEN, EDUCATION, CULTURE, AND POLITICS IN WEST GERMANY (1976); James A. Kushner, *Growth for the Twenty-First Century: Tales from Bavaria and the Vienna Woods—Comparative Images of Urban Planning in Munich, Salzburg, Vienna, and the United States*, 29 URB. LAW. 911, 924 (1997), *reprinted as modified*, 6 S. CAL. INTERDISC. L.J. 89 (1997).

362. BROWN, *supra* note 36 at 129–30; MARTIN, THE MEANING OF THE 21ST CENTURY, *supra* note 32 at 60–62.

363. FRIEDMAN, *supra* note 358 at 261–65, 288–90, 303–05.

Student. Source: U.S. Census Bureau, Public Information Office.

1. Local Response

School districts, community colleges and universities must aggressively undertake curricula and program changes to prepare students for the coming economic changes. Universities might consider specialization. Already some universities focus on agriculture and some do not include medical or law schools. Some institutions could expand or add the relevant technical training. Given the nature of the crisis, the traditional liberal arts university curriculum that offers a salad bar of different studies might be limited to a number of colleges to meet demand, but the majority of higher education institutions should be focused on what is needed to be an effective scientist, engineer, software designer, or other specialist or technician. Not only is the United States lacking technical specialists, it lacks faculty to teach mathematics and the sciences, relying on foreign-born instructors in most positions.[364] Apprenticeships should be expanded to include workers needed in the new economy, from design and construction to maintenance and servicing of trains, transit, and alternative energy systems.

Local government and local school boards should enlist local industry to establish curricula and apprenticeship programs and educate students towards the next generation of essential jobs. Schools can shift faculty towards skills, tutoring, and apprenticeship and rely more efficiently on broadcasts of virtual presentations by master teachers on cable television or podcasting. Schools can also be opened for family education to improve job skills and educate the public in sustainable living and for volunteers to teach after hours enrichment programs.

2. Regional Response

On a regional basis, schools could specialize in math, sciences, ecological sciences, computer programming, technical training, and essential skills. The development of local agriculture and food production as well as local industries to replace the globalized market of foreign goods along with new solar, wind, geothermal, wave, and other technologies should generate local jobs and satisfy local labor needs.

3. State Response

States are in a position to share tax base[365] and assure adequate education funding[366] including the funding of university and technical school education. Through a combination of scholarships and tax incentives, critical education capacity in the essential fields can be generated.

4. Federal Response

The federal government should increase funding to education, particularly through scholarships and grants, providing access for the poor and middle

---

364. *Id.*
365. ORFIELD, *supra* note 168 at 105–108; RUSK, *supra* note 168; Note, *Making Mixed-Income Communities Possible*, *supra*, note 168.
366. Robinson, *supra* note 360; Ryan, *supra* note 360.

class to the education most needed in the future. Adequate stipends to allow retraining and education with living support should speed the development in the research and development sectors, expanding the quality and impact of engineering schools, and schools teaching the sciences and construction. Foreign aid should also include aid for education, particularly involving essential technology transfer and the universities necessary to provide the specialists needed for development. Conditioning individual aid to parents that keep their children in school has been effective in both Mexico and New York City.[367]

5. International Response

International cooperation is necessary to equalize educational opportunity and to generate the trained specialists needed for emissions-free technology, efficient and sustainable food production, manufacturing of goods, and sustainable living settlements.

---

367. Kaiser, *supra* note 355.

Chapter 8

Emergency Preparedness

> In many ways, the nation's approach to preventing further terrorism in the United States and improving our capacity to respond to major disasters bears unnerving similarities to the conduct of the war in Iraq. Questionable information, misguided leadership, unclear goals, underinvestment in critical areas, and staggering hubris apply as much to large-scale disaster preparedness as they do to the conduct of the war. For this, the country has already paid a great price, with, unfortunately, more to come.
>
> IRWIN REDLENER, AMERICANS AT RISK: WHY WE ARE NOT PREPARED FOR MEGADISASTERS AND WHAT WE CAN DO NOW xxiv (2006)

> Climate change is a real threat to our national security; this time the weapons of mass destruction have been unequivocally found, not least in our own cars and power plants. If the public sector ... can fight a ferocious, years-long war based on dubious intelligence, how much more should we [be] able to do for the real thing?
>
> Lisa Heinzerling & Frank Ackerman, *Law and Economics for a Warming World*, 1 HARV. L. & POL'Y REV. 331 (2007)

The nation abounds with threatened emergencies ranging from terrorism and war with the potential of biological, chemical, or nuclear weapons to storms, pandemics, infrastructure failure, or simply rising seas from tsunamis or global warming. Preparedness must address both the instant, unannounced crises and those that gradually occur over a longer period of time. Emergency preparedness is a controversial topic in a discussion of global warming and climate change.[368] On one hand, the lessons from New Orleans counsel pre-

---

368. *See generally* ROBERT RADVANOVSKY, CRITICAL INFRASTRUCTURE: HOMELAND SECURITY AND EMERGENCY PREPAREDNESS (2006); IRWIN REDLENER, AMERICANS AT RISK: WHY WE ARE NOT PREPARED FOR MEGADISASTERS AND WHAT WE CAN DO NOW (2006) (arguing that the nation is simply unprepared); Aileen M. Marty, *Hurricane Katrina: A Deadly*

paredness in terms of governmental support and intergovernmental cooperation based upon a clear and precise disaster plan with very clear responsive assignments. Such a plan would call for clear lines of authority to handle issues, communications, and directions in time of disaster or emergency. Disaster planning for urban centers is markedly different from the planning problems presented by other populations. "The Katrina calamity presents sad corroborating evidence of what can happen when disaster strikes a densely populated urban area and its inner-city residents are not prepared or protected."[369]

Since 2004, the federal model has been the National Incident Management System (NIMS).[370] NIMS identifies a framework to organize government and private entities to cooperate to manage incidents. The Incident Command System (ICS), a component of NIMS, identifies responsibilities during incident management. Typically, leadership for managing incidents rests at the local level with an incident commander. At the state level, the governor or a designated representative is charged with coordinating state efforts and resources and communicating with incident command. At the federal level a designated agency coordinates and communicates with state and local incident command, offering assets or resources as appropriate. The failure of federal, state, and local plans are not in the plans themselves, but in the failure to accept disaster responsibility and manage incidents and, most dramatically, in the failure to adequately train to manage disasters at the various levels.[371] Despite the Katrina disaster, planning for future emergencies is not adequate; governmental agencies are still unclear on the assignment of responsibilities and rebuilding still remains a futuristic goal.[372] It has been argued that the existing state-centered framework simply does not produce the level of governmental assistance

---

Warning Mandating Improvement to the National Response to Disasters, 31 NOVA L. REV. 423 (2007) (describing shortcomings of existing system and post-Katrina modifications); Gary A. Poliakoff, *Disaster Planning and Recovery*, 31 NOVA L. REV. 457 (2007) (describing planning and plan for careful recovery).

369. Michael Greenberger, *Preparing Vulnerable populations for a Disaster: Inner-City Emergency Preparedness—Who Should Take the Lead?*, 10 DEPAUL J. HEALTH CARE L. 291 (2007) (advocating a pilot program including education and survival supplies), *available at* http://ssrn.com/abstract=1017887.

370. RADVANOVSKY, *supra* note 368 at 95–118. *See also* A LEGAL GUIDE TO HOMELAND SECURITY AND EMERGENCY MANAGEMENT FOR STATE AND LOCAL GOVERNMENTS (Ernest B. Abbott & Otto J. Hetzel eds. 2005); DEPARTMENT OF HOMELAND SECURITY, NATIONAL RESPONSE FRAMEWORK (Jan. 2008) (with supporting Annexes describing disaster management responsibilities).

371. JAMES F. MISKEL, DISASTER RESPONSE AND HOMELAND SECURITY: WHAT WORKS, WHAT DOESN'T 15–17 (2006).

372. REDLENER, *supra* note 368 at 11–15.

Superdome line, New Orleans, Aug. 28, 2005.
Source: FEMA/Marty Bahamonde.

that Americans expect or deserve in a crisis situation.[373] Rather than following the Bush Administration's idea of federalism and states' rights, the federal government needs to do more than just channel funding through the existing flawed framework for crisis management and enforce the congressional framework for federal leadership as in the cases of terrorism.[374] The Department of

---

373. E.L. Gaston, *Taking the Gloves Off of Homeland Security: Rethinking the Federal Framework for Responding to Domestic Emergencies*, 1 HARV. L. & POL'Y REV. 519, 532 (2007); Elaine C. Kamarck, *When First Responders Are Victims: Rethinking Emergency Response*, 1 HARV. L. & POL'Y REV. 97 (2007).

374. Gaston, *supra* note 373 at 524–29, 532; Kamarck, *supra* note 373 at 102–03.

Homeland Security and The Federal Emergency Management Agency (FEMA) were not prepared to respond to the catastrophic effects of Hurricane Katrina, and had varying degrees of unfamiliarity with [their] roles and responsibilities under the National Response Plan and National Incident Management System.[375] FEMA has received much criticism for relief during Hurricanes Katrina and Rita, yet by law is not directed to engage in pre-disaster activities nor to take responsibility for disaster incident management, but solely to participate in post-disaster relief and rebuilding.[376] Under the Disaster Mitigation Act of 2000, grants are available but require that state and local governments develop and submit for approval to the President a mitigation plan that outlines processes for identifying the natural hazards, risks and vulnerabilities of the area under the jurisdiction of the government.[377]

Catastrophic disaster incident management is beyond the capacity of state and local government when first responders are likely to be victims of the emergency.[378] Many experts recommend that FEMA be made a cabinet-level agency rather than be part of the Department of Homeland Security. At the cabinet level, FEMA would have enhanced power allowing disaster assessment and could make adequate plans and provide incident management, prevention, and preparation an automatic federal priority, by organizing emergency management along the lines of the military, i.e., through a regional Commander-in-Chief (CINC) command structure with authority over all federal agencies, and assure that FEMA's increasing responsibilities are not subsumed by security issues at the cabinet and agency level.[379] Rather than emergency preparedness focus on national security issues, national and domestic security should be a subdivision of an emergency preparedness agency. The Post-Katrina Emergency Management Reform Act of 2006 provides for the administrator of FEMA to be designated as a member of the cabinet by the President "in the event of national disasters, act of treason, or other man-made disasters."[380] The new

---

375. Select Bipartisan Comm. to Investigate the Preparation for and Response to Hurricane Katrina, A Failure of Initiative: Final Report, H.R. Rep. No. 109-377, at 3 (2006). *See also* John K. Pierre & Gail S. Stephenson, *After Katrina: A Critical Look at FEMA's Failure to Provide Housing for Victims of Natural Disasters*, 68 La. L. Rev. 443 (2008) (critical of FEMA's housing policies and slow response).

376. Kamarck, *supra* note 373 at 102–03.

377. Pub. L. No. 106-390, § 322 (2000), codified at 42 U.S.C. § 5165(a); 44 C.F.R. § 201 (2002). *See generally* John R. Nolan, *Disaster Mitigation Through Land Use Strategies*, 37 Envt'l L. Rptr. 10681 (2007).

378. Kamarck, *supra* note 373.

379. *Id* at 108–10.

380. Pub. L. No. 109-295, tit. VI, § 503(c)), 120 Stat. 1394 (2006).

Evacuation from Galveston for Hurricane Rita, Sept. 21, 2005.
Source: FEMA/Ed Edahl.

legislation also establishes regional FEMA offices but regional directors lack sufficient authority to focus resources on prevention and preparation for disasters as well as command during incidents, rebuilding and recovery.[381] The federal government's 2002 "Dark Winter" simulation of a terrorist attack on the U.S. demonstrated that we can anticipate absolute chaos as reliance on state and local action is "not only obsolete but downright dangerous."[382]

Following Hurricane Katrina and the mountain of blame which fell on FEMA, Congress in enacting The Post-Katrina Emergency Reform Act of 2006, sought to reform the perceived problems with FEMA. Several provisions of this attempted reorganization are particularly challenging to understand from an operational standpoint, such as those provisions which divest much of DHS's direct authority to control and reshape FEMA while still leaving FEMA as part of DHS and under the direction of the DHS Secretary.[383] Further, the

---

381. Pub. L. No. 109-295, tit. VI, §507, 120 Stat. 1394 (2006).
382. Kamarck, *supra* note 373 at 109–10.
383. *See* Keith Bea et al., Federal Emergency Management Policy Changes After Hurricane Katrina: A Summary of Statutory Provisions 7 (Dec. 15, 2006), available at http://www.fas.org/sgp/crs/homesec/RL33729.pdf.

Post-Katrina Emergency Reform Act imposed specific experience-based qualifications on presidential nominees and appointees to various FEMA positions, including the FEMA Director himself.[384] Despite concerns that these qualification requirements were unconstitutional, President George W. Bush signed the Post-Katrina Relief Act into law, binding subsequent presidents to these qualification requirements.[385]

Relocating communities and infrastructure to higher ground that would remain accessible as water levels rise, constructing levees, and developing wetlands to separate water from occupied communities seems to be still at the beginning of the talking stage. The difficulty in making specific forecasts generates skepticism, doubt, and little action. Protection against rising seas should guarantee bi-partisan support given the large population residing in the path calling for relocation of development, raising development (above ground), or establishing protection through wetlands, flood walls, and pumps. Resistance to change could be reduced by demonstrating the economic boost in construction and obtaining the support of non-coastal states by including funding for expanded reservoirs and aqueduct systems in the West and river flood protection in the Midwest.[386] In any event, the U.S. requires engineering and planning studies of every coastal city to look at the possible protections together with financial strategies that permit needed investment by states and municipalities unable to maintain existing infrastructure.[387] One city, San Diego, is using scenarios of (1) three-foot sea rise globally with a resulting tide four feet above normal and winter storms producing ten-foot wave surges, and (2) a rise of 1–3½ feet by 2100 compared to the historic rate of six inches per century.[388] In these scenarios, much of the beach areas would be wet or submerged.[389] The result and danger, should seas rise by twenty feet as some predict as early as 2015, would be obvious and would call

---

384. Pub. L. No. 109-295, tit. VI, §503(c)), 120 Stat. 1394 (2006), *codified at* 6 U.S.C. §313 (2006).

385. Alexandra R. Harrington, *Presidential Powers Revisited: an Analysis of the Constitutional Powers of the Executive and Legislative Branches over the Reorganization and Conduct of the Executive Branch*, 44 WILLAMETTE L. REV. 63, 81–82 (2007), *citing* Bea et al., *supra* note 383. Federal Emergency Management Policy Changes After Hurricane Katrina: A Summary of Statutory Provisions 7 (Dec. 15, 2006), available at http://www.fas.org/sgp/crs/homesec/RL33729.pdf.

386. Barnett & Hill, *supra* note 81.

387. *Id.*

388. *Rising Sea Levels Send Ripples Through Real Estate Industry*, SAN DIEGO UNION TRIBUNE, Jun. 24, 2007, available at http://www.signonsandiego.com/news/features/20070624-9999-lz1c24smokes.html.

389. *Id.*

for serious evacuation and building moratoria since every one-foot rise in sea level results in the ocean covering about 100 feet of dry land.[390] Although real estate brokers in the San Diego area have begun making disclosures regarding sea rise from global warming, the real estate investors and buyers appear to be indifferent.[391] Insurance companies may be unconcerned because such flooding can be excluded from insurance coverage.

On another related point, the availability of resources such as food, shelter, and medical services should be available in adequate quantity and accessible when threats occur. This principle applies to the hazards of global warming, such as floods, drought, storms, and wildfires as well as weapons of mass destruction. Special populations, such as children, the elderly, disabled, and non-English-speakers under disaster conditions require special planning and services may differ from the typical adult.[392] In 2007, The Trust for America's Health gave the federal government a D+ for post-9/11 public health emergency preparedness; more than half the states scored 5 or less out of a possible 10 points in major categories of preparedness.[393] The United States and its health care system do not have the capacity or response capabilities on a large enough scale to manage catastrophic disasters. The nation is unprepared for the possibility of a pandemic of avian flu, anthrax attacks, or the aftermath of other chemical, biological, or nuclear weapons and the same is true for potential impacts from global climate change.[394]

Each community faced with rising water or other disaster such as forest or brush fires or extreme weather must establish a reasonable plan for evacuation.[395] The most controversial issues, assuming that global warming and cli-

---

390. *Id.*
391. *Id. See generally* Gary S. Guzy, *Insurance and Climate Change*, in GLOBAL CLIMATE CHANGE AND U.S. LAW 541 (Michael B. Gerrard ed. 2007).
392. REDLENER, *supra* note 368 at 103–27.
393. *Id* at 23–24.
394. *Id* at 19–41 (2006). *See also* Ariel R. Schwartz, Note, *Doubtful Duty: Physicians' Legal Obligation to Treat During an Epidemic*, 60 STAN. L. REV. 657 (2007) (arguing that the duty to treat infectious diseases is unclear, that hospitals and their staff lack sufficient capacity and that state governors might declare an emergency and compel treatment by private physicians). *Cf.* Michael H. LeRoy, *Compulsory Labor in a National Emergency: Public Service or Involuntary Servitude? The Case of Crippled Ports*, 28 BERKELEY J. EMP. & LAB. L. 331 (2007) (interpreting existing law to permit mandatory employment of those necessary in an emergency such as dock and emergency workers and advocating legislation to protect workers from mandatory service).
395. Jonathan Jorissen, Note, *Katrina's House: The Constitutionality of the Forced Removal of Citizens from their Homes in the Wake of Natural Disasters*, 5 AVE MARIA L. REV.

Thames Barrier, London. Source: James A. Kushner.

mate change dangers persist because greenhouse gases have not been reduced significantly, relate to dealing with climate refugees, the need for replacement shelter, possible development of flood control protection, the relocation of communities from coastal areas, and the question of rebuilding. Curiously, while emergency preparedness planning responsibilities fall largely on the federal and state governments, land use planning and control is generally a completely local responsibility.[396] As a general rule, should high water result as currently forecasted, life on coastal and low-lying areas will not continue. Levees and other flood protection such as the flood gates developed in the Netherlands and in London on the Thames must be considered from a cost benefit standpoint. State and local government may be held liable for deteriorated and

---

587 (2007) (arguing for limited immunity and greater private property protection and calling for a court order for evacuation even in time of emergency, an unrealistic view when evacuations are to be ordered in anticipation for a potential catastrophe or in the face of emergency such as wildfires).

396. Nolan, *supra* note 377.

failing levees.[397] For the most part, retaining development of the coasts and permitting new development should be halted. Rebuilding flooded areas, as in the case of the Mississippi shore and New Orleans is unwise.

1. Local Response

Local government must engage in disaster planning specific to global warming in addition to other threats despite legal planning obligations resting with the state. Some all-hazard planning already in place applies to global warming, e.g., evacuation and the provision of temporary survival resources and communications. Clear leadership knowledgeable of hazards and their consequences to include those of global warming and climate change is critical. Local governments that have not done so already should negotiate and enter into mutual aid agreements with other localities to plan for interstate and intrastate cooperation during disasters as encouraged by the U.S. Federal Emergency Management Agency (FEMA).[398] Educating and encouraging families and individuals to prepare and respond will provide a community with support for change and address some of the needs that exceed the capacity of local governments. Most coastal communities must commence plans to relocate cities and settlements to higher grounds. Communities that generally have not faced flooding and a significant influx of evacuees for long term support should begin now to assess and address possible scenarios. In the Post-Katrina Emergency Management Reform Act of 2006, Congress took a step toward ensuring that it would be provided with better risk information.[399] Section 649 calls upon the administrator of FEMA to assess on an on-going basis the country's prevention capabilities.[400] But this mandate has two faults. First, the primary emphasis of sections 649–652 is on response readiness, not on prevention.[401] FEMA's primary mission is to respond to emergencies, not to assess the strength of levees or other preventative measures.[402] Second, because FEMA is an ex-

---

397. Michael J. Percy, *Delta Levees—Tort Immunity vs. Takings Liability*, 42 REAL PROP. PROB. & TR. J. 547 (2007) (arguing that takings rationale is more appropriate than blanket tort immunity when the government knowingly accepts or maintains inferior flood protection as a statewide cost saving measure).
398. MISKEL, *supra* note 371 at 124–25; RADVANOVSKY, *supra* note 368 at 79–80.
399. Pub. L. No. 109-295, tit. VI, 120 Stat. 1394 (2006).
400. *Id.* §649, 120 Stat. at 1428.
401. *Id.* §§649–652, 120 Stat. at 1428–30.
402. Fed. Emergency Mgmt. Agency, About FEMA: What We Do, http://www.fema.gov/about/what.shtm (last visited Apr. 1, 2007).

ecutive agency, it will necessarily have lower credibility within Congress than an agency controlled by Congress itself. Thus, a Congressional Risk Office would be a significant improvement. Alternatively, a separate office within the Government Accountability Office could be charged with oversight of risk issues.[403] Only local government has the authority to plan and develop disaster-resilient communities that have increased capacity to adapt to the effects of natural disasters, resulting in less property damage, environmental impact, and loss of life.[404] Just as cities are unprepared for the effects of climate change, they are unprepared for earthquakes with most affected communities needing to accomplish mandatory retrofitting of at-risk buildings and critical public and transportation facilities.[405]

2. Regional Response

At the regional level, disaster plans should include reliable and redundant communications systems with local governments and the ability to support local communities with survival resources such as medical and emergency services, food, clothing, and shelter. Planning, training, funding, and clear leadership assignments consistent with NIMS and incident command models are the critical elements for a successful regional response. Working with private industry and business partners is crucial to protect the continuation of many services, financial systems, and the economy of the areas affected since they are resources for regional planning, response, mitigation, and recovery.

3. State Response

All states must have a disaster plan that ensures prompt support of regional and local needs that includes immediate emergency services, evacuation and

---

403. Daniel A. Farber et al., *Reinventing Flood Control*, 81 Tul. L. Rev. 1085, 1123–24 (2007).

404. Nolan, *supra* note 377. *See generally* Anna K. Schwab & David Brower, *Increasing Resilience to Natural Hazards: Obstacles and Opportunities for Local Governments Under the Disaster Mitigation Act of 2000*, 38 Env't L. Rptr. 10171 (Mar. 2008); Patricia E. Salkin, *Sustainability at the Edge: The Opportunity and Responsibility of Local Governments to Most Effectively Plan for Natural Disaster Mitigation*, 38 Env't L. Rptr. 10158 (Mar. 2008).

405. Redlener, *supra* note 368 at 42–63.

relocation programs, and long-term recovery. The National Guard, a valuable asset for disaster response, is controlled by the governor, unless federalized for response. It is essential for state leadership to evaluate both the feasibility of flood control efforts and to mitigate the risk through plans to relocate residents and businesses to alternative high ground sites before disaster strikes. Some states will be more affected than others. Even those states that are not coastal states, however, will have to prepare to support interstate emergency management assistance and other mutual aid pacts. Along with climate impacts and other security-based planning, independent state seismic safety boards should assess state-wide needs relating to earthquakes and explosives that present disaster risks in addition to climate change.[406]

4. Federal Response

The primary role of the federal government is to support states' requests for emergency funding, coordinate direct support for other resources of the federal government including the active duty and reserve military to maintain security, assist in evacuation, and to provide some specialty resources and services such as medical and transportation. Under the Post-Katrina Emergency Management Reform Act of 2006, the primary responsibility to comply with federal emergency preparedness obligations is placed with the administrator of FEMA who while granted some additional powers, the authority of the position and agency are largely unchanged.[407] Unfortunately, anticipating generous disaster benefits, many private free riders will cease to insure against hazards or remove themselves from threat areas or avoid a dangerous living situation[408] and state government free riders may fail to fund disaster preparedness.[409] It has been argued that disaster relief should be left primarily to the states and local governments with the federal government providing training and planning assistance.[410] The military lacks the resources for additional missions, particularly those that have not been handled well by other agencies.[411] Nevertheless, the military has the command and control, experience, and a take charge attitude

---

406. *Id* at 6.
407. Pub. L. No. 109-295, tit. VI, §504, 120 Stat. 1394 (2006).
408. MISKEL, *supra* note 371 at 125–26.
409. *Id* at 126–29.
410. *Id* at 134–36.
411. *Id* at 137.

lacking in most federal civilian and state and local government leadership.[412] The Army Corps of Engineers, rather than serving as an agent for disaster preparedness, has been accused of constructing economically indefensible and environmentally destructive pork barrel projects for powerful politicians, and charged as the primary cause of the levee failure in New Orleans.[413] Private contractors to the states might better perform traditional state or federal responsibilities should the states not be able to function, e.g., the distribution of aid and funds, relocating refugees, providing and maintaining mass shelters, providing potable water, transporting people and supplies, and delivering emergency power.[414]

Foreign aid should include humanitarian aid to assist climate refugees, plans to accept a fair share of refugees, and aid in planning and implementing relocation and evacuation programs. Climate change refugee policy will likely eclipse the current discussion of immigration and guest worker policies and must be established before the developed world is overrun with squatter settlements. Nevertheless, each nation must consider its own ecosystems; immigration may be the largest component of population expansion and must be addressed as a hidden environmental issue.[415]

Just as the government must prepare for the health impacts of global climate change,[416] it must prepare for the unusual spread of disease, increase in climate refugees, and for climate incidents, just as it must prepare for a pandemic or a terrorist use of weapons of mass destruction. The federal government must invest in the development of vaccines for manageable and preventable diseases, i.e., smallpox or polio and immunizations or protective medications for known diseases such as influenza, reliable working plans to detect and rapidly contain early outbreaks, including the capacity of local hospitals to respond effectively, and to assure the availability of food and water for the population.[417]

---

412. REDLENER, *supra* note 368 at 160–65.
413. Michael Grunwald, *Setting the Stage for More Katrinas*, TIME.COM, Aug. 2, 2007, available at http://www.time.com/time/printout/0,8816,1649403,00.html.
414. MISKEL, *supra* note 371 at 137–38.
415. Richard D. Lamm, *Immigration: The Ultimate Environmental Issue*, 84 DENV. U. L. REV. 1003 (2007).
416. Lisa Heinzerling, *Climate Change, Human Health, and the Post-Cautionary Principle*, SSRN, Georgetown University, O'Neill Institute for National and Global Health Law Scholarship (Research Paper No. 4 Sept. 2007) (forthcoming Georgetown Law Journal), available at SSRN: http://ssrn.com/abstract=1008923 and BePress: http://1sr.nellco.org/georgetown/ois/papers/4.
417. REDLENER, *supra* note 368 at 29–41.

5. International Response

Through international cooperation, climate refugees can be resettled and provided humanitarian aid; poorer nations can be assisted in building relocation communities on high ground or flood control facilities where feasible. Similar to the National Incident Management System, nations need to organize an international incident management system to allow communication and the most efficient response and sharing of resources and assets.[418] In addition, resources are needed to provide government-guaranteed insurance by risk-averse insurers for those nations facing the dire effects of climate change in the face of inertia by developed carbon-generating nations and aid to poorer nations struck by disaster.[419]

---

418. This follows the proposal of the World Bank and the Inter-Agency Secretariat of the International Strategy for Disaster Reduction. *See* Track I, Annex C, International Strategy for Disaster Reduction (2005).

419. This follows the proposal of the Global Facility for Reduction and Recovery at the World Bank. *See* Track II, Annex D, International Strategy for Disaster Reduction (2005).

Chapter 9

Energy

> To limit [oil] dependency, the Bush Administration proposes further development of domestic sources of crude, including those in the Arctic National Wildlife Refuge (ANWR). The most optimistic estimates of this reserve's production is just under a million barrels per day— the same amount of oil that would be saved by increasing overall automobile and light-truck fuel efficiency by between two and three miles per gallon.
>
> Mark E. Eberhart, Feeding the Fire:
The Lost History & Uncertain Future of
Mankind's Energy Addiction xvii (2007)

Energy production is central to the global climate change problem. The demand for commercially traded energy has doubled between 1970 and 2005.[420] Energy consumption is expected to grow by an average of 1.1% per year between 2004 and 2030.[421] Electricity use will increase from 2007 globally by 37% by 2010 and will increase 76% by 2020.[422] China is opening a new coal-fired high carbon emission power plant large enough to serve all of Dallas, Texas, weekly[423] and intends to triple the electricity it produces from coal by 2020.[424]

---

420. Hillman et al, *supra* note 101 at 38.
421. John C. Dernbach, *Overcoming the Behavioral Impetus for Greater U.S. Energy Consumption*, 20 Global Bus. & Dev. L. J. 15, 17 (2007), *citing* U.S. Dep't of Energy, Energy Info. Admin., Annual Energy Outlook 2006 with Projections to 2030 at 136 (2006), *available at* http://www.eia.doe.gov/oiaf/aeo/index.html.
422. Rachel Oliver, *All About: Cities and Energy Consumption*, CNN.com, Dec. 31, 2007, *available at* http://www.cnn.com/2007/TECH/12/31/eco.cities/index.html.
423. Keith Bradsher & David Barboza, *Pollution From Chinese Coal Casts a Global Shadow*, N.Y. Times, Jun. 11, 2006. *See also* Mark E. Eberhart, Feeding the Fire: The Lost History & Uncertain Future of Mankind's Energy Addiction 259 (2007) (by 2035 the cumulative carbon emissions will reach a level that will be impossible to maintain a tolerable level); Martin, The Meaning of the 21st Century, *supra* note 32 at 105 (China plans to build 600 coal-fired powerplants in the near future).
424. Monbiot, *supra* note 11 at 82.

The United States is more addicted to coal than oil and its dependency on coal is greater than ever.[425] Forty percent of energy used in the United States is used to generate electricity[426] and electricity production generates 42% of the nation's carbon emissions.[427] At the end of the last century 38% of U.S. energy consumption was for building operations and the use of appliances, 36% for industrial applications and 26% for transportation.[428] Although industrial and residential use of energy is decreasing, transportation consumption has risen to 37%.[429] Lighting is the largest single source of electricity demand and accounts for 17% of total carbon emissions.[430] Unfortunately, 95% of the energy generated in the United States goes to waste in the production, transportation, and storage process.[431]

The average American's energy consumption is 20 pounds of coal daily with half a billion tons of coal burned in U.S. power plants annually.[432] One major reason for the coal industry being so prominent in the supply of energy is that it is heavily subsidized through favorable tax treatment.[433] Coal is the greatest producer of solid waste; each plant generates a million tons annually.[434] Coal is also the greatest source of overall toxic air pollution, producing 60% of it with cars and trucks producing most of the balance.[435] The burning of coal generates 41% of carbon dioxide emissions: 31% from oil and 20% from gas.[436] For each ton of coal burned, four tons of carbon dioxide are released into the at-

---

425. GOODELL, BIG COAL, *supra* note 8 at xii–xiii.
426. EBERHART, *supra* note 423 at 256.
427. Ann Brewster Weeks, *Subseabed Carbon Dioxide Sequestration as a Climate Mitigation Option for the Eastern United States: A Preliminary Assessment of Technology and Law*, 12 OCEAN & COASTAL L.J. 245, 251 (2007)
428. John J. Berger, *Renewable Energy Sources as a Response to Global Climate Concerns*, *in* CLIMATE CHANGE POLICY: A SURVEY 411, 434 (Stephen H. Schneider, Armin Rosencranz & John O. Niles eds. 2002).
429. HILLMAN ET AL, *supra* note 101 at 44.
430. *Id* at 58.
431. JAMES GUSTAVE SPETH, RED SKY AT MORNING: AMERICA AND THE CRISIS OF THE GLOBAL ENVIRONMENT 64 (2004), *citing* Robert U. Ayers, *The Energy We Overlook*, WORLD-WATCH, Nov.–Dec., 2001, at 30.
432. GOODELL, BIG COAL, *supra* note 8 at xii–xiii.
433. Roberta Mann, *Another Day Older and Deeper in Debt: How Tax Incentives Encourage Burning Coal and the Consequences for Global Warming*, 20 GLOBAL BUS. & DEV. L. J. 111, 126–27 (2007).
434. KUNSTLER, *supra* note 99 at 118.
435. *Id.*
436. FLANNERY, *supra* note 4 at 70.

mosphere.[437] Although the negative consequences of coal are known, the 2001 energy plan of the Bush Administration called for construction of 1,300 new, so-called "clean," coal-fired power plants by 2020.[438] Coal use to produce the growing demand for electricity is expected to grow by 50% between 2007 and 2020.[439]

Annually, 24,000 die prematurely in America from the effects of coal-fired power plant pollution—more deaths than AIDS, murder, or drug overdose.[440] Coal-fired power plants generate 40% of the carbon dioxide in the United States and 25% of the world's carbon dioxide emissions.[441] Coal power plants also account for 66% of all sulphur dioxide, 22% of all nitrogen oxides and 33% of all mercury emissions,[442] and account for a significant portion of other hazardous air pollutants such as lead, chromium, and arsenic.[443] In addition, coal-fired power plants generate 130 million tons of solid waste that is both combustible and contains heavy metals and other toxic compounds.[444] Frequently, the waste is pumped into abandoned mines or impoundment ponds and can leach into aquifers and water supplies.[445] The solid waste of the coal-fired power plants is three times the amount of the entire nation's municipal garbage.[446] Coal is the obvious target to slow global climate change. Unbelievably, The U.S. is going in the opposite direction. In 2025, the U.S. will burn 40% more coal than it does today.[447]

Gas-fired powerplants, although more expensive than coal, cost half as much to build, come in varying sizes thus reducing transmission costs, can be fired up or shut down quickly thereby complementing intermittent sources such as solar and wind and can also generate heat producing cogeneration.[448] Over 90% of new power generation in the U.S. today is gas-fired.[449] Natural gas use will likely double between 2007 and 2020 due to the increasing demand

---

437. *Id* at 70.
438. BROWN, *supra* note 36 at 185.
439. Oliver, *supra* note 422.
440. GOODELL, BIG COAL, *supra* note 8 at xxiv.
441. *Id* at 122.
442. *Id*.
443. *Id* at 123 (176,000 pounds of lead, 161,000 pounds of chromium, 100,000 pounds of arsenic, and 96,000 pounds of mercury).
444. *Id*.
445. *Id*.
446. *Id*.
447. MONBIOT, *supra* note 11 at 82.
448. FLANNERY, *supra* note 4 at 259–60.
449. *Id* at 260.

Geysers Geothermal Power Plant, California. Source: DOE.

for electricity.[450] If coal-fired plants on the planet were replaced by gas-fired plants, carbon emissions would be cut by 30%.[451] Only 8% of American energy comes from renewable sources; 8% is from nuclear plants; and 23% comes from coal.[452] Most of the remainder comes from oil.[453] Worldwide, however, the energy supply from wind increased 15 times between 1987 and 2007, an average growth of 30% per year.[454] Unfortunately, wind-generated energy was but 0.5% of the global electrical supply in 2004.[455]

Currently, the planet is consuming 20% more biomass (all living things) than is sustainable.[456] In 1986, humans reached the Earth's carrying capacity and since that time the earth's population has been running the environmen-

---

450. Oliver, *supra* note 422.
451. FLANNERY, *supra* note 4 at 260.
452. HILLMAN ET AL, *supra* note 101 at 41–42.
453. *Id.*
454. GEO$_4$, *supra* note 19 at 49.
455. *Id.*
456. FLANNERY, *supra* note 4 at 78.

tal equivalent of a deficit budget, sustained by plundering its capital base.⁴⁵⁷ Should the nation fail to convert to renewable energy from fossil fuels, biomass in the form of wood will be needed for heating and cooking causing extraordinary deforestation.⁴⁵⁸ Between 1800 and 1980, humans produced and consumed 244 petajoules (a thousand million million joules) of energy but between 1980 and 1999, production and consumption increased to 117 petajoules, nearly half the total of the preceding 180 years.⁴⁵⁹ By 2001, the biomass deficit had reached 20% and Earth's population was at 6 billion.⁴⁶⁰ As the population grew by 140% between 1950 and 2000, fossil fuel consumption increased nearly 400%.⁴⁶¹ Estimates of increased energy demand vary from 57% by 2020, 60% from 2020 to 2030 and 100% (or double) by 2030.⁴⁶² By 2020, the world population will expand by 20% to 8.1 billion.⁴⁶³ Using these projections, if humans can be found in 2050, they will require the equivalent of two planets' worth of resources to support their power needs.⁴⁶⁴ Global energy demand is expected to increase by 60% between 2002 and 2030, while carbon emissions are projected to increase by 62%.⁴⁶⁵ Another estimate would double world energy consumption between 2000 and 2030, most of which is projected to come from fossil fuels.⁴⁶⁶ By 2050, the growth in world population and the increase in demand for energy will be double the already excessive demand of 2007. At the rate that new coal-fired power plants are being developed, the point of inability to stabilize warming is forecast to occur in 2017.⁴⁶⁷ If proposed power plants are built, 570 billion additional tons of coal will be burned, or the equivalent of all coal burned in the last 250 years and the possibility of stabilizing the climate will be zero.⁴⁶⁸

The emphasis on nonrenewable energy is a function of worldwide annual subsidies of $700 billion for fossil fuel burning.⁴⁶⁹ In the United States, the fed-

---

457. *Id* at 78–9.
458. KUNSTLER, *supra* note 99 at 138.
459. FLANNERY, *supra* note 4 at 79.
460. *Id.*
461. GOODELL, BIG COAL, *supra* note 8 at xiv.
462. DILIP HIRO, BLOOD OF THE EARTH: THE BATTLE FOR THE WORLD'S VANISHING OIL RESOURCES 331 (2007).
463. *Id.*
464. FLANNERY, *supra* note 4 at 79.
465. INNOVATION IN ENERGY TECHNOLOGY: COMPARING NATIONAL INNOVATION SYSTEMS AT THE SECTORAL LEVEL 13 (2006).
466. GOODELL, BIG COAL, *supra* note 8 at xvi.
467. *Id* at 207.
468. *Id.*
469. BROWN, *supra* note 36 at 233.

eral government spends $20 billion annually to subsidize the development of oil, coal, and natural gas, and more than $10 billion to subsidize nuclear energy.[470] Federal and state subsidies and tax benefits to the oil industry amount to as much as $113 billion annually.[471] By comparison, annual federal oil subsidies and tax benefits exceed the total cost of the September 11, 2001 terrorist attacks on the World Trade Center.[472] When the social costs of oil, including health costs, crop losses, damage to buildings, materials, and forests, and water pollution are added with government subsidies, the military costs of defending oil supplies excluded, the total bill may be as high as $806.9 billion and possibly $1 trillion annually.[473] The subsidies alone amount to $2,700 per person and if eliminated, would result in gas prices increasing by $6 per gallon and save in subsidies the equivalent of the cost to provide health insurance for the 45 million uninsured and build 1,500 new schools in every state.[474]

Were energy generated from sources not emitting pollution, the public health savings would be $53 million per year.[475] At the same time that the U.S. government subsidizes fossil-fuel industries that help cause global warming, it provides less than $1 billion to subsidize alternative renewable energy that could help in avoiding global warming.[476] Low, subsidized electric rates encourage environmentally harmful metal smelting operations such as those at gold mines and encourage excessive consumption.[477] A small portion of the current subsidies that should be shifted to renewable energy sources and technology could be used to retrain or buy out coal miners and other fossil fuel workers.[478]

The shift to alternative renewable energy sources[479] has been impressive in a few European countries but has been largely symbolic in the United States.

---

470. GELBSPAN, BOILING POINT, *supra* note 39 at 185; GELBSPAN, THE HEAT IS ON, *supra* note 11 at 180.
471. TAMMINEN, *supra* note 148 at 60–61.
472. *Id* at 61.
473. *Id* at 61–62.
474. *Id* at 61.
475. Mark J. Spalding & Charlotte de Fontaubert, *Conflict Resolution for Addressing Climate Change With Ocean-Altering Projects*, 37 ENV'T L. RPTR. 10740 (2007).
476. MARTIN, THE MEANING OF THE 21ST CENTURY, *supra* note 32 at 47 ($20 billion annual subsidies for fossil fuel).
477. BROWN, *supra* note 36 at 243.
478. GELBSPAN, BOILING POINT, *supra* note 39 at 185.
479. *See generally* RENEWABLE ENERGY: POWER FOR A SUSTAINABLE FUTURE (2d ed. Godfrey Boyle ed. 2004); RENEWABLE RESOURCES AND RENEWABLE ENERGY: A GLOBAL CHALLENGE (Mauro Graziani & Paolo Fornasiero eds. 2007).

Although the current laws offer tax credits and other support valued at $14.5 billion for renewable energy investment, subsidies for fossil fuels are often hidden, such as renewable electricity production credits provided to a facility that burns coal using "Indian coal" (coal owned by certain tribes) or a combination of coal and biomass.[480] Current tax law offers a renewable electricity production credit, tax credit-supported energy bonds, a business solar investment tax credit, a business fuel cell credit, and a residential energy-efficient property credit.[481] Conservation incentives include energy property credits, deductions for energy-efficient improvements in commercial buildings, energy-efficient new home credits, energy-efficient appliance credits, and preferential tax treatment for alternative fuel vehicles.[482] Incentives for renewable energy and conservation, however, are overshadowed by huge incentives for fossil fuel use, including drilling and development costs, depletion allowances, credits for production from nonconventional sources, oil recovery credit, renewable energy production credit—including from coal, clean renewable energy bonds covering coal facilities co-fired with biomass or Indian coal, and tax credits for investing in so-called "clean coal facilities."[483] Efforts to address the negative environmental and health consequences of burning coal led to he Clean Coal Power Initiative Act that has provided $1.8 billion in subsidies to corporations to develop the oxymoron "clean coal."[484] The transportation of the coal to be cleansed will require vastly more petroleum than the theoretical process could ever save.[485] Coal burning as well as coal mining is simply dangerous and environmentally destructive.[486] Nevertheless, processes that decarbonize coal or natural gas have been developed and may when economically feasible and in operation offer additional cleaner interim sources of energy.[487] Significant re-

---

480. 26 U.S.C. §§ 45(d)(2)(ii), 45(c)(9) (2005); Roberta Mann, *Subsidies, Tax Policy, and Technological Innovation*, in Global Climate Change and U.S. Law 565, 567–69 (Michael B. Gerrard ed. 2007).

481. 26 U.S.C. § 45 (2005); Mann, *Subsidies, Tax Policy, and Technological Innovation*, in Global Climate Change and U.S. Law, *supra* note 480 at 567–74.

482. Mann, *Subsidies, Tax Policy, and Technological Innovation*, in Global Climate Change and U.S. Law, *supra* note 480 at 572–76.

483. *Id* at 576–79.

484. 42 U.S.C.A. §§ 15961–64 (2007). *See generally* Edwin Black, Internal Combustion: How Corporations and Governments Addicted the World and Derailed the Alternatives 278–79 (2006); Mann, *Subsidies, Tax Policy, and Technological Innovation*, in Global Climate Change and U.S. Law, *supra* note 480.

485. Black, *supra* note 484 at 278.

486. *Id* at 278–79.

487. Hiro, *supra* note 462 at 270–71.

ductions in carbon emissions can come from the use of co-generation plants that generate heat and power by burning pellets generated from recycling in a system that emits no carbon.[488]

Proposals to utilize technology to convert coal into liquid fuels is not promising as the process produces more than twice the greenhouse gas emissions as gasoline.[489] Hydroelectric power could be increased as an alternative to fossil-fuel energy through upgrading and retrofitting existing dams as only 3% of existing dams produce electricity and as many as 2,600 existing dams can be retrofitted.[490] New dams and hydroplants cost 100 times the cost of retrofit.[491] Yet, dams are controversial as they prevent the spawning of fish and may threaten wilderness areas.[492] More than 460 dams have been destroyed in the last decade with 2.5 million still standing.[493] The Chinese development of the Three Gorges Dam will generate 18,000 megawatts, the equivalent of burning 40 million tons of coal each year, and despite environmental harm, may be the only way to avoid global warming.[494] England is considering a barrage or enclosure dam across the Severn Estuary with 200 large turbines that would provide 7% of the entire country's electricity—the equivalent of 12 nuclear power stations.[495]

As of 2005, there were 104 nuclear reactors providing 20% of the U.S. electric supply.[496] Nuclear power, although it involves lowered emissions, is not feasible without subsidies because the payment of the costs of development, waste disposal, accident insurance, the cost of decommissioning worn out plants, and their potential as a terrorist or foreign target renders nuclear power unattractive.[497] In addition, only 5 to 25 years' supply of uranium exists.[498] Alter-

---

488. SPETH, *supra* note 431 at 65.

489. *The Coal Trap*, N.Y. TIMES, May 30, 2007 (editorial also cites M.I.T. estimate that it will cost $70 billion to build enough coal-to-liquid plants to replace 10 percent of U.S. gasoline consumption).

490. DIANE RAINES WARD, WATER WARS: DROUGHT, FLOOD, FOLLY, AND THE POLITICS OF THIRST 146–47(2002).

491. *Id* at 147.

492. *Id* at 148–9.

493. *Id* at 148.

494. *Id* at 149.

495. *Id* at 150.

496. HILLMAN ET AL, *supra* note 101 at 101.

497. BROWN, *supra* note 36 at 39–40.

498. DAVID GOODSTEIN, OUT OF GAS: THE END OF THE AGE OF OIL 104–110 (2004). *See also* CRAIG MORRIS, ENERGY SWITCH: PROVEN SOLUTIONS FOR A RENEWABLE FUTURE 80 (2006) (although current reserves might last 50 to 60 years, if fossil-fuel-generated electricity were shifted to nuclear, economically viable reserves would last three to four years;

native materials are too close to weapons grade and alternative processes involve yet undeveloped technologies.[499] To convert from coal-fired power to nuclear by 2020 would require opening a 1,000 megawatt nuclear plant every 1.61 days at a cost 35% higher than investing in energy-efficient technology.[500] Nuclear power is also not carbon-emission free; each plant generates 230,000 tons of carbon dioxide over its life.[501] Nuclear powerplants are also not dependable as a shortage of fresh water may require shut downs.[502] However, to avoid a dramatic drop in lifestyle with the termination of inexpensive fossil fuel, nuclear may be the sole choice if the nation continues to fail in aggressively pursuing renewable energy choices.[503]

Solar power presents extraordinary potential as solar energy falling on the United States alone amounts to 10,000 times as much electric power as is consumed in the country.[504] Each day, more solar energy falls on the earth than the entire planet would use in 27 years; only 1% of that energy is harnessed.[505] Solar is the most promising energy technology.[506] On the local level, solar domestic hot-water collectors are affordable, efficient and can be paid off in ten years, replacing the second most expensive element of home utilities after heating.[507] Europe is planning to create a $10 billion solar project in the Jordanian desert that will generate one-sixth of the European Union's electricity needs.[508]

Wind is another extremely promising source of energy. Wind power from just North Dakota, Kansas, and Texas alone could satisfy the entire energy

---

by 2025, Europe use of nuclear at 23% in 1995, will drop to 9% by 2025 and 1% by 2035). *But see* KUNSTLER, *supra* note 99 at 141 (estimating perhaps 100 years of uranium).

499. GOODSTEIN, *supra* note 498 at 80.

500. BRENDA VALE & ROBERT VALE, GREEN ARCHITECTURE: DESIGN FOR AN ENERGY-CONSCIOUS FUTURE 52 (1991).

501. *Id* at 53.

502. Mitch Weiss, *Drought Could Shut Down Nuclear Plants*, ASSOCIATED PRESS, Jan. 23, 2008.

503. KUNSTLER, *supra* note 99 at 140–46; Goodell, *The Prophet of Climate Change*, *supra* note 70.

504. GOODSTEIN, *supra* note 498 at 40.

505. ROGERS & KOSTIGEN, *supra* note 209 at 137.

506. Bernow, *in* CLIMATE CHANGE POLICY 189, 194–95, *supra* note 93. *See also* BLACK, *supra* note 484 at 284–85.

507. PAHL, *supra* note 99 at 33–37.

508. Robin McKie, *How Africa's Desert Sun Can Bring Europe Power*, THE OBSERVER, Dec. 2, 2007, *available at* http://www.guardian.co.uk/environment/2007/dec/02/renewableenergy.solarpower/.

Heliostats at the Solar Two Power Plant, Daggett, California. Source: DOE.

needs of the nation.[509] A farmer can receive $3,000 to $5,000 in annual royalties for siting a single wind turbine on a quarter-acre of land, a trade-off of only $120 from 40 bushels of corn or $15 from beef.[510] Erecting a wind generator can cut the farmer's dependency on carbon-generating electricity.[511] For an investment of $13,000, an Illinois family installed a 56-foot wind turbine

---

509. BROWN, *supra* note 36 at 188, *citing* D.L. ELLIOT ET AL, AN ASSESSMENT OF THE AVAILABLE WINDY LAND AREA AND WIND ENERGY POTENTIAL IN THE CONTIGUOUS UNITED STATES (1991). *See generally* PAUL GIPE, WIND POWER: RENEWABLE ENERGY FOR HOME, FARM, AND BUSINESS (2004); PAHL, *supra* note 99 at 63–103; Howard E. Susman & Kathleen J. Doll, *Wind Advisory: Finding a Suitable Site for a Wind Farm Requires More Than Locating a Blustery Location*, 30 LOS ANGELES LAW. 35 (Jan. 2008).

510. BROWN, *supra* note 36 at 191.

511. Jackson, *supra* note 96 (reporting that a farmer in Iowa constructed a wind turbine for $140,000 less a federal government grant of $29,000 that produces twice the energy used and could sell the excess or bank it with the power company for when there is no wind with an investment payoff of 10 to 15 years which shorten should electricity rates increase as anticipated while generating no carbon emissions).

Flowind, Altamont Pass, California. Source: DOE.

and reduced their monthly electricity bill from $90 to $10.[512] A 10-kilowatt turbine with average wind speed of 12 miles per hour can reduce carbon emissions equivalent to removing 1.3 cars from the road.[513] Early turbines, such as those in Altamont, California, produced 55 kilowatts. Today, 2-megawatts (2,000 kilowatts) are now available and 5-megawatt turbines are under development.[514] The cost of wind energy has dropped from 30 cents a kilowatt hour to 5 cents.[515] Unfortunately, wind energy development has been slow due to Congress creating uncertainty by not consistently maintaining tax credits.[516]

Solar energy is more promising than wind-generated energy because wind-generated energy can only grow to 25% of the national power source without

---

512. Kristina Shevory, *Homespun Electricity, From the Wind*, N.Y. TIMES, Dec. 13, 2007.
513. *Id.*
514. MORRIS, ENERGY SWITCH, *supra* note 498 at 127. *See also* BLACK, *supra* note 484 at 283–84.
515. MORRIS, ENERGY SWITCH, *supra* note 498 at 127. *See also* BLACK, *supra* note 484 at 283–84.
516. BROWN, *supra* note 36 at 189.

requiring costly modification of the power grid.[517] Geothermal power is also an efficient and promising alternative to fossil fuels.[518] Another promising technology is the use of tidal turbines that, unlike dams, allow free flowing water to generate energy.[519] Biomass, including ethanol, however, is not a promising energy replacement as it requires an underlying fossil fuel platform.[520] Ethanol from corn generates twice the carbon emissions of gasoline when the land use changes needed for increased corn production are considered.[521] It takes 1.29 gallons of petroleum to produce one gallon of ethanol.[522] In comparison to corn and food-based ethanol, biofuels such as cellulosic or sugar-based ethanol could reduce greenhouse gas emissions.[523] While sugar cane-based ethanol would cost $35 a barrel, as compared to oil which is anticipated to reach $105 per barrel by the end of 2008, corn costs $81 a barrel, wheat $145, soybeans $232, and cellulose $305.[524] The diversion of agricultural land to biofuel is causing a rapid rise in food costs and famine.[525] At least one study suggests that carbon emissions can be reduced 12% by production and combustion of ethanol and 41% by biodiesel.[526] Unfortunately, coal-fired carbon-generating plants are being built to produce ethanol because it is inexpensive and allows an investor to take advantage of federal subsidies.[527] Brownfields can also make excellent sites for renewable energy such as biomass or a wind farm at a

---

517. MORRIS, ENERGY SWITCH, *supra* note 498 at 144.

518. BLACK, *supra* note 484 at 280–83; MORRIS, ENERGY SWITCH, *supra* note 498 at 133–39.

519. Michael B. Walsh, Comment, *A Rising Tide in Renewable Energy: The Future of Tidal In-Stream Energy Conversion (TISEC)*, 19 VILL. ENVTL. L.J. 193 (2008) (arguing that the technology is commercially viable and carries little environmental impact).

520. KUNSTLER, *supra* note 99 at 138.

521. Associated Press, *Study: Corn Ethanol No Climate Solution: Greenhouse Emissions Much Higher if Land Use Factored In, Researchers Say*, MSNBC.COM, Feb. 7, 2008, available at http://www.msnbc.msn.com/id/23057867/.

522. BLACK, *supra* note 484 at 286–87.

523. KUNSTLER, *supra* note 99 at 140–46.

524. Ambrose Evans-Pritchard, *Why the Price of 'Peak Oil' is Famine*, TELEGRAPH.CO.UK, Feb. 9, 2008, *available at* http://www.telegraph.co.uk/money/main.jhtml?xml=/money/2008/02/07/cnoil107.xml.

525. *Id.*

526. Dernbach, *U.S. Policy*, in GLOBAL CLIMATE CHANGE AND U.S. LAW *supra* note 185 at 61, 71, *citing* Jason Hill et al., *Environmental, Economic, and Energetic Costs and Benefits of Biodiesel and Ethanol Biofuels*, 103 PROC. NAT'L ACAD. SCI. 11,206 (2006).

527. PAHL, *supra* note 99 at 191–92.

landfill.[528] Waste vegetable oil biofuel is an extremely efficient and inexpensive source of power for small-scale agriculture.[529] Vegetable oil, including used cooking oils can serve as a modest partial replacement for oil.

Converting to renewable energy from the fossil fuel era will not only dramatically reduce pollution, greenhouse gases, and global climate change, it would unleash an economic and jobs bonanza for the economy.[530] The strategy necessary now requires full, war-time like development of renewable energy sources together with a similar effort at conservation and reduced energy demand in transportation, buildings, and appliances. The Manhattan Project that developed the atomic bomb cost $1.89 billion.[531] By comparison, the government spends $31.5 billion on ordinary bombs, mines, and grenades; $24 billion on small arms; and $64 billion on tanks annually; $20 billion annually could generate energy independence.[532]

Conservation can have dramatic effects on energy demand. For example, more efficient household appliances would eliminate the need for 127 new power plants.[533] More efficient air conditioning would eliminate the need for 43 new power plants.[534] Raising commercial air conditioning standards would eliminate the need for 50 new power plants.[535] Using tax credits and energy codes to improve new building efficiency would save another 170 plants.[536] Retrofitting existing buildings to higher energy standards would eliminate 210 plants.[537] Heating and air conditioning demand in new buildings can be reduced as much as 90% by modern insulation, triple-glazed windows with tight seals, and passive solar design.[538] Conservation would also be advanced by eliminating incandescent light bulbs (20% of electricity is used for lighting), shifting to hybrid cars, and redesigning urban transport systems to raise efficiency and increase mobility.[539] Shifting from incandescent light bulbs to com-

---

528. Steven Ferrey, *Converting Brownfield Environmental Negatives into Energy Positives*, 34 B.C. ENVTL. AFF. L. REV. 417 (2007).
529. PAHL, *supra* note 99 at 218–23.
530. GOLDSTEIN, *supra* note 314.
531. BLACK, *supra* note 484 at 276.
532. *Id.*
533. BROWN, *supra* note 36 at 185, *citing* Bill Prindle, *How Energy Efficiency Can Turn 1300 New Power Plants Into 170, fact sheet* (May 2, 2001).
534. BROWN, *supra* note 36 at 185, *citing* Prindle, *supra* note 533.
535. BROWN, *supra* note 36 at 185, *citing* Prindle, *supra* note 533.
536. BROWN, *supra* note 36 at 185 (2006), *citing* Prindle, *supra* note 533.
537. BROWN, *supra* note 36 at 185, *citing* Prindle, *supra* note 533.
538. SPETH, *supra* note 431 at 65.
539. BROWN, *supra* note 36 at 185–86.

pact fluorescent lamps (CFLs) would also yield a risk-free investment returning between 25 to 40% each year.[540] CFLs save four-fifths the electricity and cause one-fifth the pollution as tungsten bulbs.[541] The most serious concern with renewable energy alternatives is that from a pragmatic standard, the shift cannot be made quickly enough, e.g., small solar projects and small wind turbines are not of a sufficient scale to replace the current overwhelming dependence on fossil fuels.[542]

1. Local Response

Local governments might establish overlay ordinances, i.e., single ordinances that allow regulations to be imposed on a wide geographical area that includes many different zoning districts and add to the existing regulation, to permit wind[543] and other renewable energy facilities in any district. Local energy providers can choose to invest and acquire power from renewable producers. Calgary, Canada's public power company, has erected an 80 megawatt wind farm to supply municipal buildings and civic operations with 75% of energy needs.[544] Local government should be a showplace for alternative energy within its vehicle fleets, buildings, and in conditions placed upon contractors. In addition, the building code can require green construction and should impose limits on the size of buildings and dwellings. Taxes might be imposed on buildings or dwellings that exceed a reasonable minimum such as 400 square feet per occupant. Climate change impacts and carbon dioxide and other greenhouse gas emissions should be significant criteria in all state and local permit procedures including those involving energy facilities.[545]

Al Gore outlined a number of personal strategies to reduce energy use, including (1) energy-efficient lighting, (2) energy-efficient appliances, (3) proper

---

540. *Id* at 186. *See also* GOLDSTEIN, *supra* note 314 at 65–68 (compact fluorescent lamps last five or more years and are cleaner than traditional light bulbs).
541. VALE & VALE, *supra* note 500 at 45.
542. TERTZAKIAN, *supra* note 99 at 206–07.
543. *See* Mark K. Dausch, Comment, *Analyzing a Municipality's Authority to Enact the Model Ordinance for Wind Energy Facilities in Pennsylvania*, 45 DUQ. L. REV. 47 (2006); Susman & Doll, *supra* note 509.
544. Chris Turner, *The Secret Greening of Calgary*, GLOBE & MAIL, Sept. 15, 2007.
545. Laura H. Kosloff & Mark C. Trexler, *Consideration of Climate Change in Facility Permitting*, in GLOBAL CLIMATE CHANGE AND U.S. LAW 259 (Michael B. Gerrard ed. 2007); Edna Sussman, *Reshaping Municipal and County Laws to Foster Green Building, Energy Efficiency, and Renewable Energy*, 16 N.Y.U. ENVTL. L.J. 1 (2008); (green building and alternative energy encouragement).

operation and maintenance of appliances, (4) efficiently heat and cool homes using programable thermostats to reduce energy use while sleeping or at work, (5) home insulation, (6) obtaining a smart energy audit, (7) conserve hot water, (8) reduce standby power waste of electronic equipment and adaptors, (9) improve efficiency of home office and computer operation by favoring laptops, and (10) switch to green power.[546]

2. Regional Response

The regional response to energy change will be to assure regional planning to provide land for renewable energy development and production and to include renewable energy in all of government's activities. Allowing electric customers a choice of electricity source such as renewable energy and imposing emissions caps and reduction standards and enforcement could be a regional obligation.[547]

3. State Response

The state and federal legislative response to climate change has been has been multi-faceted albeit an ineffective attempt at mitigation.[548] Carbon taxes would encourage emission reductions from motor vehicles, building, and energy generation.[549] Such a carbon tax would be imposed on suppliers of coal,

---

546. GORE, *supra* note 4 at 306–10; Mann, *Another Day Older and Deeper in Debt*, *supra* note 433 at 136–141.

547. Eleanor Stein, *Regional Initiatives to Reduce Greenhouse Gas Emissions*, in GLOBAL CLIMATE CHANGE AND U.S. LAW 315 (Michael B. Gerrard ed. 2007).

548. *See* Alice Kaswan, *The Domestic Response to Global Climate Change: What Role for Federal, State, and Litigation Initiatives?*, 42 U.S.F.L. REV. 39 (2007) (reviewing existing and proposed legislative responses).

549. Dallas Burtraw & Paul R. Portney, *A Carbon Tax to Reduce the Deficit*, in NEW APPROACHES ON ENERGY AND THE ENVIRONMENT: POLICY ADVICE FOR THE PRESIDENT 19 (Richard D. Morgenstern & Paul R. Portney eds. 2004) (recommending gradual pacing of increases so as not to destroy the value of existing long-lived investments); Goulder & Nadreau, *in* CLIMATE CHANGE POLICY, *supra* note 342 at 120; Ian W.H. Parry, *Fiscal Interactions and the Case for Carbon Taxes Over Grandfathered Carbon Permits*, in CLIMATE-CHANGE POLICY 218 (Dieter Helm ed. 2005). *See also* Claudia Dias Soares, *Critical Issues in Implementing Energy Taxation, in* EU CLIMATE CHANGE POLICY: THE CHALLENGE OF NEW REGULATORY INITIATIVES 256–73 (Marjan Peeters & Kurt Deketelaere eds. 2006); *Quebec to Collect Nation's 1st Carbon Tax: Energy Companies will Pass Cost to Consumers, Say Analysts*, CBC NEWS, Jun. 7, 2007) (reporting a tax of 0.8 cents on each litre of gas and 0.9

crude oil, and natural gas based upon the carbon content of each.[550] Both federal standards and state power regulation can require the development of renewable energy such as solar, wind, hydro, bio-mass, bio-fuel, nuclear, or alternative technologies.[551] Although ideally power generation should be carbon-free, current power plants can have emissions capped and new power plants can be required to offset a percentage of emissions as well as reducing emissions,[552] investing in projects that offset emissions, or making payments into a renewable energy trust fund.[553] California has imposed a mandatory 25% emissions reduction program setting a cap based on 1990 levels and has established a mandatory emissions reporting system.[554] California's S.B. 1368 prohibits in-state power distributors from buying power under a long-term contract (defined as any contract for longer than five years) with an out-of-state power generator that fails to comply with emissions standards set by the California Energy Commission.[555] The standards adopted require that greenhouse gas emissions be set according to natural gas-based power generation In this way, the legislation is tailored to target current "dirty" coal production, without banning coal permanently, given the possibility for development of cleaner, if not clean, coal technology.[556] It can be argued that despite the burden on out-of-state energy producers, the courts should defer to the state legislature and sustain the legislation over any Commerce Clause challenge.[557]

---

cents on each litre of diesel fuel expected to generate $200 million annually to support energy-saving initiatives and public transport).

550. Goulder & Nadreau, *in* CLIMATE CHANGE POLICY, *supra* note 342 at 131.

551. HOUGHTON, *supra* note 13 at 203–21. *See also* PITTOCK, *supra* note 36 at 170–83; RENEWABLE ENERGY: POWER FOR A SUSTAINABLE FUTURE (2d ed. Godfrey Boyle ed. 2004).

552. Robert B. McKinstry, Jr., *Laboratories for Local Solutions for Global Problems: State, Local and Private Leadership in Developing Strategies to Mitigate the Causes and Effects of Climate Change*, 12 PENN ST. ENVTL. L. REV. 15, 35–39 (2004) (Massachusetts and New Hampshire programs described).

553. Steven Ferrey, *Sustainable Energy, Environmental Policy, and States' Rights: Discerning the Energy Future Through the Eye of the Dormant Commerce Clause*, 12 N.Y.U. ENVTL. L.J. 507 (2004) (describing Oregon program).

554. California Global Warming Solutions Act of 2006, *available at* http://www.leginfo.ca.gov/cgi-bin/postquery?bill_number=ab_32&sess=PREV&house=B&author=nunez. For a general discussion of state initiatives to reduce carbon emissions, see James A. Holtkamp, *Dealing with Climate Change in the United States: The Non-Federal Response*, 27 J. LAND RESOURCES & ENVTL. L. 79 (2007).

555. Act of Sept. 29, 2006, ch. 598, §2, 2006 Cal. Legis. Serv. at 3795; Nordberg, *supra* note 166 at 2074–75.

556. Nordberg, *supra* note 166 at 2074–75.

557. Id.

States can also offer tax incentives for development, investment in alternatives such as solar tax credits, and consumer investment in carbon-free and carbon-reduced mechanical systems and appliances and the retrofitting of homes. For example, Germany provided landowners tax incentives to install solar and other renewable power and provided for the sale of excess power to the local utility at double the regular electricity price.[558] Many states have passed legislation encouraging renewable energy and facilitating participation by small local producers in the power grid.[559] Wisconsin has pursued a policy of requiring emissions reporting that can have positive effects but found that purely voluntary incentive programs were insufficient.[560] States can also enter into regional interstate compacts to cooperatively reduce carbon emissions in energy production.[561]

Legislatures in 22 states and the District of Columbia impose renewable energy portfolio standards requiring utilities to generate some energy from renewable sources.[562] California has imposed obligations of mandatory conservation, en-

---

558. MORRIS, ENERGY SWITCH, *supra* note 498 at 105–118.

559. Steven Ferrey, *Nothing But Net: Renewable Energy and the Environment, MidAmerican Legal Fictions, and Supremacy Doctrine*, 14 Duke Envtl. L. & Pol'y F. 1 (2003).

560. BARRY G. RABE, STATEHOUSE AND GREENHOUSE: THE EMERGING POLITICS OF AMERICAN CLIMATE CHANGE POLICY 93–106 (2004).

561. Note, *The Compact Clause and the Regional Greenhouse Gas Initiative*, 120 HARV. L. REV. 1958 (2007).

562. BARRY G. RABE, PEW CTR. ON GLOBAL CLIMATE CHANGE, RACE TO THE TOP: THE EXPANDING ROLE OF U.S. STATE RENEWABLE PORTFOLIO STANDARDS 3–4 (2006), *available at* http://www.pewclimate.org/docUploads/RPSReportFinal%2Epdf. *See* CAL. PUB. UTIL. CODE § 399.11 (West Supp. 2008) (20% by 2010); COLO. REV. STAT. ANN. § 40-2-124 (West 2007) (3% to 2007, 5% to 2010, 10% to 2014, 15% to 2019 and 20% after 2020); CONN. GEN. STAT. § 16-245a (2007) (5% after 2006, 10% by 2010 with one additional percent annually with 23% after 2019); Del. Code Ann. Tit. 26, § 354 (2006) (10% by 2019); Haw. Rev. Stat. § 269-92 (Supp. 2006) (10% by 2011, 20% by 2021); 20 ILL. COMP. STAT. ANN. 687 (West Supp. 2007); Response to Governor's Sustainable Energy Plan for the State of Illinois, Ill. Commerce Comm'n 05-0437 (July 19, 2005), *available at* http://www.epa.gov/cleanenergy/documents/gta/guide_action_chap6_s1.pdf (mandatory 1% increase annually until 2013 and wind must be 75% of renewables); Me. Rev. Stat. Ann. 35A § 3210(3)-A (2007) (1% by 2008, 10% by 2016); MD. CODE ANN., PUB. UTIL. COS. § 7-703 (Supp. 2007); MASS. GEN. LAWS ANN. ch. 25A, § 11F (West 2002); 225 MASS CODE REGS. 14.07 (2007) (1% in 2003 to 4% by 2009); MINN. STAT. ANN. § 216B.1691 (West Supp. 2008) (12% by 2012, 17% by 2016, 20% by 2020, and 25% of state's power from renewable energy by 2025 and the state's largest energy supplier must generate 30% from renewables by 2020); MONT. CODE ANN. § 69-3-2004 (2007) (5% of retail sales for 2008 through 2009; 10% 2010 through 2014; 15% for 2015 and successive years enforced through fines but exempting cooperatives); NEV. REV. STAT. ANN § 704.7821 (LexisNexis Supp. 2005) (6% for 2005–2006, increasing by 3% every two years until it reaches 20% in 2015, with 5% each year reflecting

ergy efficiency, renewable energy, research, and development by utility companies.[563] States have also sought to reduce carbon emissions by establishing caps on carbon emissions,[564] including emissions caps on existing[565] and new[566] power plants, or requiring electricity producers to pay a fee per ton to offset emissions.[567] Idaho has imposed a moratorium on new coal-fired power plants.[568] Illinois has mandated municipal carbon emissions reductions.[569] Eleven states have adopted California's[570] restrictive tailpipe emissions rules for motor vehicles.[571]

---

solar power); N.J. Admin. Code tit. 14, §8-2.3 (2008); N.M. Admin. Code tit. 17.9.572.10 (2008) (5% by 2006 increasing 1% annually to 10% by 2011, 15% generated by 2015 and 20% by 2020 assuming cost reasonable); N.Y. Pub. Serv. Comm'n, Case No. 03-E-0188 (Sept. 24, 2004), *available at* http://www3.dps.state.ny.us/pscweb/WebFileRoom.nsf/0/85D8CCC6A42DB86F85256F1900533518/$File/301.03e0188.RPS.pdf?OpenElement (from current level of 19% to 25% by 2013); 73 Pa. Cons. Stat. Ann §§ 1648.1–.8 (West Supp. 2007) (18% by 2021); R.I. Gen. Laws § 39-26-4 (2006) (3% before 2008, 12% before 2020); Tex. Util Code Ann. § 39.904 (Vernon 2007) (5,880 megawatts by 2015, 500 of those other than wind, with 10,000 megawatt goal for 2025); Wis. Stat. Ann. § 196.378 (West 2002 & Supp. 2007) (1.55% by 2008, 2.2% by 2012, and 10% by 2015). *See generally* David Hodas, *State Initiatives*, in Global Climate Change and U.S. Law 343, 355–59 (Michael B. Gerrard ed. 2007).

563. Cal. Pub. Res. Code § 25751 (West 2007 & Supp. 2008).

564. Cal. Health & Safety Code §§ 38500–38599 (West Supp. 2008) (Global Warming Solutions Act) (goal only); N.H. Rev. Stat. Ann. § 125-O:3 (Supp. 2007).

565. Cal. Health & Safety Code § 38562 (West Supp. 2008) (mandatory caps beginning 2012); 310 Mass. Code Regs. § 7.29 (2008); N.H. Rev. Stat. Ann. §§ 125-O:3 to -O:4 (Supp. 2007) (allowing reduction or purchase of emissions credits).

566. Or. Rev. Stat. § 469.501 (2005) (authority to adopt); Or. Admin R. 345-024-0550 (2007); Wash. Rev. Code Ann. § 80.70.20 (West Supp. 2008) (carbon dioxide emissions mitigation plan).

567. Or. Admin R. 345-024-0560 (2008).

568. Idaho Code Ann. §§ 39-124, 39-125 (Supp. 2007) (two-year but exempting those regulated by PUC).

569. 415 Ill. Comp. Stat. 145/1 (2008) (Illinois permission for signs designating "cool cities" where reduce emissions by 7% from 1990 levels by 2012).

570. Cal. Health & Safety Code § 43018.5 (West 2006 & Supp. 2008); 13 Cal. Code Regs. § 1961.1 (2008).

571. Ariz. Rev. Stat. Ann. § 49-542 (Supp. 2007); Conn. Agencies Regs. § 22a-174-36b (2005); Me. Code R. § 06-096, ch. 127 § 2 (2008); 310 Mass. Code Regs. 7.40 (2005); 7 N.J. Admin. Code § 27B-5.3 (2008); 13 N.J. Admin. Code §§ 20-43.1, 43.21 (2008); N.Y. Comp. Code R. & Regs. tit. 6, §§ 210, 218 (2005); Or. Rev. Stat. § 803.350(8)(a) (2007); 75 Pa. Cons. Stat. Ann. § 4706(c)) (West 2006); R.I. Gen. Laws. §§ 31-47.1-1 to 47.1-11 (2002); Vt. Code R. § 12-031-001 (2007); Wash. Rev. Code Ann. §§ 70.120A.010 (West Supp. 2008) (adopting but rejecting adoption of California zero emission vehicle standards and conditions adoption on Oregon's adoption for particular model years).

A number of states have imposed mandatory efficiency standards for state motor fleets, including vehicles, tires, and fuels.[572] Some states have established a public benefit fund raised from a fee or surcharge on electricity rates used to invest in clean energy supplies.[573] Alternative energy investment has also been encouraged through policies such as net metering, i.e., allowing retail electric customers who generate their own electricity from renewable sources to sell excess electricity back to the grid,[574] together with grants,[575]

---

572. Cal. Health & Safety Code §43018.5 (West 2006); Haw. Rev. Stat. §103D-412 (Supp. 2006) (20% of new state vehicles to be hybrid or alternative-fuel with percentage increasing over time); Iowa Exec. Order No. 41 (2005) (all vehicles hybrids or alternative fuels by 2010); S.C. Exec. Order No. 2001-35 (2001), *available at* http://www.scstatehouse.net/archives/executiveorders/exor0135.htm (state agencies operating alternative fuel vehicles required to use alternative fuels if practical and economically feasible).

573. Cal. Pub. Res. Code §§25740–25751 (West 2007 & Supp. 2008); Haw. Rev. Stat. §269-121 to 269-124 (Supp. 2006). *See generally* Hodas, in Global Climate Change and U.S. Law, *supra* note 562 at 359.

574. Ark. Code Ann. §§23-18-601 to 23-18-604 (2002 & Supp. 2007); Cal. Pub. Util. Code §2827 (West Supp. 2008); Colo. Rev. Stat. §§40-9.5-301 to 40-9.5-305 (2007); 4 Colo. Code Regs. §723-3, Rule 3664 (2007); Ga. Code Ann. §§46-3-50 to 46-3-56 (2004); Haw. Rev. Stat. §§269-101 to 269-111 (Supp. 2006); 170 Ind. Admin. Code R. 4-4.2 (2008); Kan. Stat. Ann. §66-1,184 (2002); Ky. Rev. Stat. Ann. §§278.465 to 278.468 (LexisNexis Supp. 2007); http://yosemite.epa.gov/OAR/globalwarming.nsf/UniqueKeyLookup/SHSU5BUTYL/$File/netmetering.pdf (Maine program established by public utilities commission); Md. Code Ann., Pub. Util. Cos. §7-306 (LexisNexis Supp. 2007); Mo. Ann. Stat. §386.887 (West 2007); Mo. Code Regs. Ann. tit. 4, §240-20.065 (2007); Mont. Code Ann. §§69-8-601 to 69-8-605 (2007); Nev. Rev. Stat. Ann. §704.768 (NexisLexis 2003); N.H. Rev. Stat. §362-A:9 (Supp. 2007); Ohio Rev. Code Ann. §4928.67 (West 2000); Or. Rev. Stat. §757.300 (2005); 73 Pa. Stat. §§1648.2, 1648.5 (West 2007); Utah Code Ann. §§54-15-102, 54-15-103 (Supp. 2007); Vt. Stat. Ann. tit. 30, §219a (Supp. 2007); Wash. Rev. Code Ann. §§80.60.005 to 80.60.040 (West 2001 & Supp. 2008); Wyo. Stat. Ann. §§37-16-101 to 37-16-104 (2007). Hodas, in Global Climate Change and U.S. Law, *supra* note 562 at 362–63.

575. Pace Law School Center for Environmental Legal Studies, *State Response to Climate Change: 50-State Survey*, in Global Climate Change and U.S. Law 371, 372 (Michael B. Gerrard ed. 2007) (Alabama to governmental agencies); Ark. Code Ann. §15-4-2804 (2003); Fla. Stat. Ann. §§377.805 to 377.806 (West Supp. 2008) (rebates for photovoltaic and solar thermal technology installations on commercial buildings and matching grants for research and development of alternative energy vehicles and technologies); 2006 Haw. Sess. Laws 096, Pt. 2, §2, *available at* http://www.capitol.hawaii.gov/session2006/bills/HB2175_cd1_.htm (H.B. 2175 appropriated $5 million for solar power in schools); 20 Ill. Comp. Stat. Ann. 687/6-3, 687/6-4 (West Supp. 2007); 220 Ill. Comp. Stat. Ann. 5/16-111.1 (West 2007); Iowa Code Ann. §266.39C (West 2003); Me. Rev. Stat. Ann tit. 35-A. §3211-C (Supp. 2007) (solar rebate program); Md. Code Ann., State Gov't §9-2007 (Supp. 2007) (solar powered heating); Minn. Stat. Ann. §216B.241 (West Supp. 2008) (solar power genera-

loans,[576] and tax credits[577] and exemption of alternative energy investment

---

tion rebates up to $2,000); MINN. STAT. ANN. §216C.41 (West Supp. 2008) (wind power generation); N.H. REV. STAT. ANN. §362-F:10 (West Supp. 2007); N.M. STAT. ANN. §§6-21D-1 to 6-21D-10 (2005) (tax-exempt bonds for renewable energy improvements to schools and state buildings); N.M. STAT. ANN. §§68-2-29 to 68-2-33 (2002) (tree planting for sequestration); N.M. STAT. ANN. §71-7-6 (2007); Pace Law School Center for Environmental Legal Studies, *supra* at 394 (New York), see www.nyserda.org (last visited Feb. 13, 2008); N.D. CENT. CODE §54-44.5-09 (Supp. 2007) (wind to hydrogen project); 73 Pa. Cons. Stat. §1647.3 (West. Supp. 2007) (alternative fuels and vehicles for government agencies and profit and non-profit organizations funding from gross receipts tax on electric utilities); R.I. GEN. LAWS §39-2-1.2(b) (2006) (funded by charge to utility customers); VA. CODE ANN. §45.1-392 (2002) (photovoltaic); WASH. REV. CODE ANN. §70.94.960 (West 2002) (clean fuel vehicles for local government and school buses); WIS. STAT. ANN. §16.957 (West 2003).
576. Pace Law School Center for Environmental Legal Studies, in GLOBAL CLIMATE CHANGE AND U.S. LAW, *supra* note 575 at 372 (Alabama to private entities and governmental agencies for interest on loans); ALASKA STAT. §§45.88.010 to 45.88.140 (2006) (to governmental units and cooperatives); 20 ILL. COMP. STAT. ANN. 687/6-4 (West Supp. 2007); IOWA CODE ANN. §476.46 (West Supp. 2007) (half of each loan at no interest for 20 years); MINN. STAT. ANN. §41B.046 (West Supp. 2008) (wind or anaerobic-digestion power generation for farmers); MISS. CODE ANN. §57-39-39 (West 1999); MO. ANN. STAT. §§640.651 to 640.686 (West 2006); MONT. CODE ANN. §§75-25-101, 80-12-201 (2007); NEB. REV. STAT. §§66-1001 to 66-1011 (2003); OHIO REV. CODE ANN. §§4928.61 to 4928.63 (Supp. 2007); OR. REV. STAT. §§470.050 to 470.210 (2005) (fuels and vehicle modification); TENN. CODE ANN. §4-3-710 (2005) (small business).
577. ARIZ. REV. STAT. §41-1510.01 (2006); ARIZ. REV. STAT. §43-1083 (Supp. 2007) (25% of cost up to $1,000 for solar or wind in residences); ARK. CODE ANN. §§15-4-2104, 15-4-2801 to 15-4-2805 (2003) (tax refund and credits); COLO. REV. STAT. ANN. §39-22-516 (2007) (hybrid vehicles); FLA. STAT. ANN. §§220.192, 220.193 (West Supp. 2008); Ga. Code Ann. §§48-7-40.16 (2005) (10 to 20% of purchase or lease of zero or low-emission vehicle); HAW. REV. STAT. ANN. §235-12.5 (Supp. 2006) (35% of purchase price and installation costs of solar systems and 20% for wind systems); IOWA CODE ANN. §476C (West Supp. 2007) (to providers for renewable power sold to customers); LA. REV. STAT. ANN. §§47:38, 47:287.757 (2001) (20% of cost of converting vehicle to alternative fuel or purchasing new vehicle); ME. REV. STAT. ANN. tit. 36, §§5219-P, 5219-AA (Supp. 2007); MD. CODE ANN., TAX-GEN. §10-722 (Supp. 2004) (6 to 8% for green buildings, 20 to 25% for photo voltaic generators, 20% for wind turbines, 30% for fuel cells); MICH. COMP. LAWS ANN. §208.39e (West 2003) (including tax credits for payroll of alternative energy companies); Mo. ANN. STAT. §§135.300 to 135.311 (West 2000) (to those who process wood for fuel); MONT. CODE ANN. §§15-32-115, 15-32-401 (2007); N.M. STAT. ANN. §7-2A-19 (2007); N.M. STAT. §7-9G-2 (2007) (30% income tax credit for residents to install photo-voltaic and solar thermal systems); Pace Law School Center for Environmental Legal Studies, in GLOBAL CLIMATE CHANGE AND U.S. LAW, *supra* note 575 at 394 (New York), see www.nyserda.org (last visited Feb. 13, 2008); N.D. CENT. CODE §57-38-01.8 (Supp. 2007) (3% for five years); OKLA. STAT. ANN. tit. 68, §2357.32A (West Supp. 2008); OR. REV. STAT. §§316.116, 469.165 to 469.170 (2005) (alternative fuel

from property,[578] sales and use, or excise taxes,[579] replacement generation[580] or transaction tax.[581] Maine bans fees for energy-generating customers for reduction, elimination, or re-establishment of electric service where customers have alternative energy supplies.[582]

Some states have imposed obligations of state government to construct buildings using green technology and to generate a portion of the electricity that they consume.[583] A number of states have established green building stan-

---

and vehicle investment); R.I. Gen. Laws §§ 44-57.1 to 44-57.12 (2005) (25%); S.C Code Ann. § 12-6-3587 (2007) (20% of federal credit for purchase of clean vehicle and solar heating and cooling and landfill gas); Utah Code Ann. § 59-10-1009 (Supp. 2007).

578. Cal. Rev. & Tax Code § 73 (West Supp. 2008); Kan. Stat. Ann. § 79-201h (1997); La. Rev. Stat. Ann. § 47:1706 (West 2006) (solar power used for buildings and swimming pools); Mich. Comp. Laws Ann. § 211.9i (West 2005) (personal property exemption for alternate energy companies); Minn. Stat. Ann. §§ 272.02, 272.028 (West 2007); Mont. Code Ann. §§ 15-6-225, 15-24-1401 (2007); Nev. Rev. Stat. § 361.0687 (2007); N.D. Cent. Code § 57-02-08(27) (Supp. 2007) (5-year solar and wind exemption); Or. Rev. Stat. § 307.175 (2007); R.I. Gen. Laws § 44-3-21 (2005) (enabling ordinances for local exemption); S.D. Codified Law §§ 10-6-35.8 to 10-6-35.20 (alternative energy exemption), 10-4-36 to 38 (2004) (wind production assessed at local rather than state level); Tenn. Code Ann. § 67-5-601 (2006) (wind); W. Va. Code Ann. § 11-6A-5a (LexisNexis 2003 & Supp. 2007) (wind turbines valued at salvage value); Wis. Stat. Ann. § 70.111(18) (West Supp. 2005).

579. Ariz. Rev. Stat. §§ 42-5061, 42-5075(B)(14) (Supp. 2007) (up to $5,000 for solar); Fla. Stat. Ann. § 212.08 (West Supp. 2008) (solar power generators); Iowa Code Ann. § 423.3 (West Supp. 2007); Md. Code Ann., Tax-Gen. § 11-207 (2004) (residential heating); Minn. Stat. Ann. §§ 297A.67 (solar power generation), 297A.68 (West 2007) (wind power generation); N.D. Cent. Code §§ 57-39.2-04, 57-43.2-02 (Supp. 2007) (hydrogen for fuel cells, internal combustion engines, and storage facilities and wind generator facility owners); R.I. Gen. Laws § 44-18-40.1 (2005); S.C Code Ann. § 12-36-2110 (2000 & Supp. 2007) (tax on manufactured homes meeting energy-efficiency standards capped at $300); Utah Code Ann. § 59-12-104 (2006); Vt. Stat. Ann. tit. 32, § 9741(46) (2001 & Supp. 2007); Wyo. Stat. Ann. § 39-15-105(a)(viii)(N) (2007) (excise tax exemption until 2012).

580. Iowa Code Ann. § 437A.6 (West 2006).

581. Iowa Code Ann. § 437A.6 (West 2006) (wind power exempt from replacement generation tax); Mont. Code Ann. § 15-72-104 (2007) (on state or reservation lands for renewable energy generated).

582. Me. Rev. Stat. Ann. tit. 35-A, § 3209(3) (2007).

583. Ariz. Exec. Order No. 2005-05 (2005), http://www.governor.state.az.us/eo/2005_05.pdf (10%); Cal. Exec. Order No. S-20-04 (2004), *available at* http://www.dot.ca.gov/hq/energy/ExecOrderS-20-04.htm (state buildings must reduce energy use by 20% by 2015); Cal. Gov't Code § 14684 (West Supp. 2008) (state buildings to install solar energy systems by 2007); 20 Ill. Comp. Stat. Ann. 3105/10.04 (West Supp. 2007); Iowa Exec. Order No. 41 (2005), *available at* http://www.dsireusa.org/documents/Incentives/IA08R.pdf (10%); Minn. Stat. Ann. § 16B.32 (West 2005).

dards,[584] while Idaho[585] and Wisconsin[586] have mandated the use of green building codes. Nevada has pre-empted municipal housing and building codes to provide permission for green buildings.[587] All green buildings can be provided priority processing.[588] Hawaii[589] and Nevada[590] have banned zoning or covenants restricting solar or wind power installation.[591] Other states have established a right to solar access as a property right based on prior appropriation to allow an unobstructed line of sight from a solar collector to the sun.[592] Oregon exempts tenant-installed alternative generation equipment from fixture status thereby allowing removal by the tenant at the end of the lease.[593]

Conservation measures undertaken by states include the establishment of household appliance efficiency standards.[594] Several states allow hybrid-pow-

---

584. Haw. Rev. Stat. §§ 46-19.6, 196-9 (Supp. 2006); 20 Ill. Comp. Stat. 3953/25 (2007) (requiring use of Energy Star-labeled light bulbs in state buildings); Tex. Gov't Code Ann. § 2166.403 (Vernon Supp. 2007) (energy alternative in new and existing buildings if feasible); Vt. Exec. Order No. 14-03 (2003), *available at* http://governor.vermont.gov/tools/index.php?topic=ExecutiveOrders&id=249&v=Article; Wis. Stat. Ann. § 13.48, 101.027 (West Supp. 2007) (higher energy-efficient standards in building codes); Wis. Exec. Order No. 145 (2006), *available at* http://www.wisgov.state.wi.us/journal_media_detail.asp?prid=1907 (standards based on LEED certification to be adopted and energy use by state buildings reduced by 10% by 2008, 20% by 2010).
585. Idaho Code Ann. §§ 39-4109, 39-4116 (Supp. 2007).
586. Wis. Stat. Ann. § 13.48, 101.027 (West Supp. 2007) (higher energy-efficient standards in building codes).
587. Nev. Rev. Stat. § 278.580 (2002 & Supp. 2005) (straw and other renewable sources for construction and code-compliant solar or wind energy systems to power buildings).
588. Haw. Rev. Stat. Ann. § 46-19.6 (Supp. 2006).
589. Haw. Rev. Stat. Ann. § 196-7 (Supp. 2006).
590. Nev. Rev. Stat. §§ 111.239, 278.0208 (NexisLexis 2005) (solar and wind).
591. *See also* Susman & Doll, *supra* note 509.
592. N.H. Rev. Stat. Ann. §§ 477:49 to 477:51 (2001); N.M. Stat. Ann. §§ 47-3-1 to 47-3-5 (1995).
593. Or. Rev. Stat. § 90.265 (2007).
594. Ariz. Rev. Stat. Ann. §§ 1375 to 1375.03 (Supp. 2007) (12 appliances); Cal. Admin. Code §§ 1601 to 1608 (2008); California Energy Comm'n, Appliance Efficiency & Appliance Regulations, http://www.energy.ca.gov/efficiency/appliances/ (last visited Feb. 24, 2006) (17 appliances); Conn. Gen. Stat. Ann § 16a-48 (West 2007); Pace Law School Center for Environmental Legal Studies, in Global Climate Change and U.S. Law, *supra* note 575 at 386 (Maryland established standards for nine appliances); N.J. Rev. Stat. §§ 48:3-99 to 48:3-106 (2005); N.Y. Energy Law Appx. Pt. 7825.1 to 7826.1 (McKinney 2004); R.I. Gen. Laws §§ 39-27-1 to 33-27-9 (2006); Vt. Stat. Ann. §§ 9-2791 to 9-2798 (2006); Wash. Rev. Code Ann. §§ 19.260.010 to 19.260.900 (West 2007) (13 appliances). *See generally* Hodas, in Global Climate Change and U.S. Law, *supra* note 562 at 363–65.

ered motor vehicles to be driven without passengers in high occupancy vehicle lanes.[595] Maine permits low speed vehicles such as golf carts to be operated on public streets.[596] Oregon has exempted electric and natural-gas-driven vehicles from compliance with pollution control system requirements.[597] A number of states have imposed mandatory conservation requirements on state agencies.[598]

4. Federal Response

The state and federal legislative response to climate change has been multifaceted albeit an ineffective attempt at mitigation.[599] All subsidies and tax incentives for fossil fuels should be eliminated thereby generating a significant public dividend.[600] Instead, investment tax credits should be offered to encourage development of renewable energy.[601] The Bureau of Land Management should implement a program of generating wind power on federal lands[602] and even consider Denmark's example of making every family a share owner in its wind-power industry.[603] The United States government should cease its funding of rural electrification through loans to build coal-fired

---

595. FLA. STAT. ANN. §316.0741 (West 2006); GA. CODE ANN. §32-9-4 (2006) (conditioned on federal approval).

596. ME. REV. STAT. ANN. tit. 29-A, §102 (Supp. 2007).

597. Or. Rev. Stat. §§815.295 to 815.300, 468A.365 (2005).

598. HAW. REV. STAT. §§46-19.6, 196-9 (Supp. 2006) (20% of new state vehicles to be hybrids or alternative fuel); IOWA EXEC. ORDER NO. 41 (2005), *available at* http://www.dsireusa.org/documents/Incentives/IA08R.pdf (reduce 15% of energy use from 2000 by 2010 and 20% of bulk diesel fuel to be renewable); ME. REV. STAT. ANN. tit. 5, §1812E (West 2002) (fuel efficient vehicles of at least 45 miles per gallon for cars and 35 for light trucks); WIS. STAT. ANN. §16.75 (West Supp. 2007) (mandatory purchase of 20% of energy needs from renewables by six largest agencies); Wis. Exec. Order No. 145 (2006), *available at* http://www.wisgov.state.wi.us/journal_media_detail.asp?prid=1907(energy use by state buildings reduced by 10% by 2008, 20% by 2010).

599. *See* Kaswan, *supra* note 548 (reviewing existing and proposed legislative responses). For a discussion of a normative executive order increasing energy efficiency and reducing greenhouse gases, see Charles Openchowski, *The Next Greenhouse Gas Executive Order?*, 38 ENVTL. L. REP. NEWS & ANALYSIS 10077 (2008).

600. Berger, *in* CLIMATE CHANGE POLICY 411, 440, *supra* note 428.

601. *Id.*

602. Roy Fuller, *Wind Energy Development on BLM Lands*, 24 J. LAND RESOURCES & ENVTL. L. 613 (2004); Gregory M. Adams, Comment, *Bringing Green Power to the Public Lands: The Bureau of Land Management's Authority and Discretion to Regulate Wind-Energy Developments*, 21 J. ENVTL. L. & LITIG. 445 (2006).

603. 3 JOHANSEN, *supra* note 1 at xiii.

power plants.⁶⁰⁴ Foreign aid and investment should be targeted to renewable energy production and technology. The Energy Policy Act of 2005 requires 7.5 billion gallons of ethanol be used as a transportation fuel by 2012.⁶⁰⁵ President Bush announced a "Twenty in Ten" plan, in his 2007 State of the Union address that seeks to reduce gasoline consumption by 20% by 2017, relying primarily on ethanol with a goal of 35 billion gallons in use by 2017.⁶⁰⁶ Corn and other food-based ethanol subsidies should be eliminated.

5. International Response

Emission reduction treaties should be established that reduce and discourage motor vehicles and fossil fuel energy generation such as the Kyoto Protocol⁶⁰⁷ that calls for reducing 1990 emissions by a set percentage.⁶⁰⁸ Although the Bush Administration and the Republican Congress refused to support Kyoto, 400 cities in 50 states have officially ratified and endorsed the Kyoto Protocol.⁶⁰⁹ Even so, the goals of the Kyoto Protocol are grossly inadequate in light

---

604. *U.S. Loans for Coal Plants Clash with Carbon Cuts*, WASH. POST.COM, May 14, 2007 (reporting $35 billion in loans under New Deal program aimed at rural poverty) (available on msnbc).
605. Energy Policy Act of 2005, § 1501, 42 U.S.C.A. § 7545(o)(2)(B)(I) (2007) (amending the Clean Air Act to establish a renewable fuel program).
606. President George W. Bush, State of the Union Address (Jan. 23, 2007), *available at* http://www.whitehouse.gov/news/releases/2007/01/20070123-2.html); Twenty in Ten: Strengthening America's Energy Security, *available at* http://www.whitehouse.gov/stateoftheunion/2007/initiatives/energy.html.
607. Marc Pallemaerts & Rhiannon Williams, *Climate Change: The International and European Policy Framework*, in EU CLIMATE CHANGE POLICY: THE CHALLENGE OF NEW REGULATORY INITIATIVES 22, 37–41 (Marjan Peeters & Kurt Deketelaere eds. 2006) (summarizing and analyzing components). *See also* CLIMATE CHANGE AND THE KYOTO PROTOCOL: THE ROLE OF INSTITUTIONS AND INSTRUMENTS TO CONTROL GLOBAL CHANGE (Michael Faure *et al.* eds. 2003); FLEXIBILITY IN CLIMATE POLICY: MAKING THE KYOTO MECHANISMS WORK (Tim Jackson *et al.* eds. 2001); Christoph Böhringer & Michael Finus, *The Kyoto Protocol: Success or Failure?*, in CLIMATE-CHANGE POLICY 254, 280–81 (Dieter Helm ed. 2005) (despite its ostensible failure to reduce carbon emissions, the mechanism is a positive start towards an effective control strategy); Zoya E. Bailey, Comment, *The Sink that Sank the Hague: A Comment on the Kyoto Protocol*, 16 TEMP. INT'L & COMP. L.J. 103 (2002).
608. Goulder & Nadreau, *in* CLIMATE CHANGE POLICY, *supra* note 342 at 120; MONBIOT, *supra* note 11 at 48 (Kyoto calls for but a 5.2% reduction by 2012).
609. James Boyce, *The Greening of Politics: Seven Years of Rapid Change*, MSN.COM, Sept. 1, 2007, *available at* http://stopglobalwarming.msn.com/article.aspx?&cp-documentid=5288548.

of current increases in emissions and forecasts of devastation by warming.[610] Cutting current emissions by half, however, is well within range if aggressive steps are taken.[611] The thrust of Kyoto is to make carbon emissions reductions a domestic policy, thus leaving strategies up to each nation.[612]

Emission permits can be established to allow those who generate carbon to purchase credits allowing emissions in exchange for the credit-sellers reducing emissions or not reaching their permitted emissions cap.[613] For example, under the Kyoto Protocol, Russia is below its 1990 carbon level and would be able to sell excess capacity permits.[614] The Kyoto Protocol approach is likely to result in carbon leakage whereby restricted nations outsource greenhouse-intensive activities to a nation unfettered by the protocol.[615] Kyoto, despite a political focus on the failure of support by the United States, is simply too much of a compromise and it is insufficient to halt and reverse global climate change.[616] The administration of President George W. Bush has been charged with inactivity and even sabotage of efforts to prevent global warming, however, the Clinton-Gore administration has been charged with destroying the Kyoto Protocol.[617] Kyoto calls for a 5.2% carbon dioxide emissions reduction when scientific consensus calls for a 60–80% reduction to halt global warming.[618] Although trade sanctions appear to be a viable incentive for international carbon emissions compliance, so far the Kyoto Protocol has been neither efficient

---

610. BROWN, *supra* note 36 at 182–83.
611. *Id* at 184–85.
612. Chris Backes & Reinske Teuben, *Legal Aspects of the Dutch Approach to CO_2 in* CLIMATE CHANGE AND THE KYOTO PROTOCOL: THE ROLE OF INSTITUTIONS AND INSTRUMENTS TO CONTROL GLOBAL CHANGE 128, 129 (Michael Faure *et al.* eds. 2003).
613. Goulder & Nadreau, *in* CLIMATE CHANGE POLICY, *supra* note 342 at 120–31; MacDonald & Makuch, *in* EU CLIMATE CHANGE POLICY, *supra* note 342 at 125–48; Peeters, *in* EU CLIMATE CHANGE POLICY, *supra* note 342 at 169–86; Pring, *in* EU CLIMATE CHANGE POLICY, *supra* note 116 at 188–201; Donehower, *supra* note 116 (arguing that without the incorporation of China and the United States, the world's two largest polluters, the carbon markets may serve as a successful market tool and example of the efficiency of an open market to cost-efficiently solve environmental problems, but will do nothing to curb GHG emissions and limit the effects of climate change).
614. Wybe Th. Douma, *The European Union, Russia and the Kyoto Protocol, in* EU CLIMATE CHANGE POLICY: THE CHALLENGE OF NEW REGULATORY INITIATIVES 51, 62–64 (Marjan Peeters & Kurt Deketelaere eds. 2006).
615. HENSON, *supra* note 29 at 39.
616. 3 JOHANSEN, *supra* note 1 at xxiii.
617. MONBIOT, *supra* note 11 at v.
618. *Id* at 48; SMITH, *supra* note 108 at 15–16.

nor effective and it has been argued that restricting trade of goods on the basis of ratification or compliance would likely be unwise.[619]

Carbon taxes would encourage emission reductions from motor vehicles, building, and energy generation.[620] Such a carbon tax would be imposed on suppliers of coal, crude oil, and natural gas based upon the carbon content of each.[621] Were such a tax imposed internationally, the scheme would eliminate the need for a system of trading carbon permits.[622] A hybrid of taxing and permit trading might consist of setting national carbon goals and allowing the trading of carbon permits subject to a cap on permit prices.[623] Should countries such as the United States believe that global warming is not a catastrophic threat, they should be willing to establish an insurance fund that would compensate nations facing the most catastrophic consequences anticipated.[624] International cooperation must set effective emissions standards and assist poorer nations to improve economically and environmentally. The expansion of carbon-generating energy production is more dangerous to the planet's population than weapons of mass destruction. Sanctions are justified against nations that continue to develop or fail to modify and eliminate their carbon generation. It has been argued that carbon emissions must be addressed at the international level and cannot be effectively addressed at the local level.[625]

---

619. Julian Morris, *Warming Aid, Chilling Trade?*, in ADAPT OR DIE: THE SCIENCE, POLITICS AND ECONOMICS OF CLIMATE CHANGE 133, 151 (Kendra Okonski ed. 2003).

620. Burtraw & Portney, in NEW APPROACHES ON ENERGY AND THE ENVIRONMENT, *supra* note 549 at 19 (recommending gradual pasing of increases so as not to destroy the value of existing long-lived investments); Goulder & Nadreau, in CLIMATE CHANGE POLICY, *supra* note 342 at 120. See also Parry, in CLIMATE-CHANGE POLICY, *supra* note 549 at 218; Soares, in EU CLIMATE CHANGE POLICY, *supra* note 549 at 256–73.

621. Goulder & Nadreau, in CLIMATE CHANGE POLICY, *supra* note 342 at 131.

622. *Id.*

623. *Id* at 134–36; Donehower, *supra* note 116 (arguing that without the incorporation of China and the United States, the world's two largest polluters, the carbon markets may serve as a successful market tool and example of the efficiency of an open market to cost-efficiently solve environmental problems, but will do nothing to curb GHG emissions and limit the effects of climate change).

624. Loucks, in CLIMATE CHANGE POLICY, *supra* note 55 at 511.

625. Jonathan B. Wiener, *Think Globally, Act Globally: The Limits of Local Climate Policies*, 155 U. PA. L. REV. 1961 (2007).

Chapter 10

Housing and Construction

> In building or renovating their homes for energy efficiency, what can Americans learn from the European successes of the last ten years? While air conditioning accounts for about a fifth of the power consumption in the US, central and northern European homes rarely have air conditioning at all. Thus, in large parts of the US, passive houses will have to make air conditioners redundant. Luckily, increasing shading and using geothermal power to cool indoor air will take us a long way. However, poorly designed technology will not help. My parents have an energy-efficient air conditioning and hot water system in their new home in southern Mississippi but unfortunately the hot water tank is in the garage at the other end of the house from their shower. The water may be heated efficiently, but in the time it takes for the hot water to get to the master bedroom, I could be finished showering. Had the architect thought to put the hot water tank where the master bedroom is rather than where the cars are, the savings would be greater and hot water would be provided faster.
>
> It would be a great step forward if we could begin to understand sustainability as a combination of new technology and common sense.
>
> CRAIG MORRIS, ENERGY SWITCH: POWER SOLUTIONS FOR A RENEWABLE FUTURE 178–79 (2006)

Housing policy has a great impact on carbon emissions. Construction can include green architecture or the use of materials that require less energy in production, such as recyclables rather than wood or steel,[626] and green site

---

626. *See generally* PETER BUCHANAN, TEN SHADES OF GREEN: ARCHITECTURE AND THE NATURAL WORLD (2005); BUILDING WITHOUT BORDERS: SUSTAINABLE CONSTRUCTION FOR THE GLOBAL VILLAGE (Joseph F. Kennedy ed. 2004) (emphasis on straw); DEAN HAWKES & WAYNE FORSTER, ENERGY EFFICIENT BUILDINGS: ARCHITECTURE, ENGINEERING, AND ENVIRONMENT (2002); SMITH, *supra* note 108; JAMES STEELE, ECOLOGICAL ARCHITECTURE: A CRITICAL HISTORY (2005); VALE & VALE, *supra* note 500; Jason R. Busch et al., *Tax and Financial Incentives for Green Building*, 30 Los Angeles Law. 15 (Jan. 2008); Benjamin S. Kings-

planning.[627] Passive solar heating and cooling through a heat exchange process involving water circulation through a ground piping system that heats in winter and cools in summer; south-facing, super-insulated, triple-glazed windows; organic or garden roofs, and limited external surfaces all require lower amounts of energy.[628] Insulation made from recycled newspapers can significantly reduce energy consumption as can energy-saving systems and appliances. Solar or wind power can also be integrated into the production of houses and other buildings. For example, the technology exists to provide buildings with heating and air-conditioning that uses passive solar techniques rather than electricity but the solar industry is undercapitalized and the government is not supporting development.[629]

Reuse of existing structures, particularly when made energy-efficient, will result in significant reduction of emissions as compared to demolition and new construction. It can dramatically reduce solid waste; Building demolition generates half of all land fill and other solid waste. Building a new home with recycled wood can save 3.2 acres of forest.[630] Often, older homes are built using wood from old-growth forests which is four times more energy efficient than new growth lumber and will last 3 to 4 times as long.[631] Through 2050, 89 million new or replacement homes and 190 billion square feet of new offices, institutions, stores, and other non-residential buildings will be constructed.[632] Since two-thirds of all buildings projected to be constructed in the United States by 2050 have not yet been built, a great opportunity exists to reduce emissions

---

ley, *Making it Easy to be Green: Using Impact Fees to Encourage Green Building*, 83 N.Y.U. L. Rev. 1 (2008) (draft version) (advocating fees levied on non-green development, using proceeds to subsidize LEED certified gold and platinum buildings together with exactions measured by the projected use of energy and infrastructure such as water, sewer, and roads); Charles J. Kibert & Kevin Grosskopf, *Envisioning Next-Generation Green Buildings*, 23 J. Land Use & Envtl. L. 145 (2007); Stephen T. Del Percio, Student Article, *The Skyscraper, Green Design & LEED Green Building Rating System: The Creation of Uniform Sustainable Standards for the 21st Century or the Perpetuation of an Architectural Fiction?*, 28 Environs 117 (2004).

627. American Society of Landscape Architects, The Sustainable Sites Initiative—Standards and Guidelines: Preliminary Report (2007), *available at* http://greenerbuildings.com/news_detail.cfm?NewsID=36194&print=true.

628. Monbiot, *supra* note 11 at 59–78.

629. Elizabeth Douglass, *His Passion for Solar Still Burns*, L.A. Times, Nov. 10, 2007.

630. Rogers & Kostigen, *supra* note 209 at 139.

631. Id.

632. Reid Ewing et al., Growing Cooler: The Evidence On Urban Development and Climate Change 8 (2007).

Bio-01 sustainable community (West Harbor), Mälmo, Sweden.
Source: James A. Kushner.

and increase sustainability by changing building techniques and construction materials.[633]

Planting two shade trees on the west side of the home and one on the east side can save 20% of the air-conditioning bill and if shade trees were planted around 25% of air-conditioned dwellings, the energy saved would be enough to shut down three coal-fired power plants.[634] Miami, Florida found that electricity bills were 10% lower in areas with more than 20% tree cover as compared to treeless areas.[635] One tree can release 400 litres of water daily which can cool the air.[636] If one in ten rural households planted a windbreak of evergreen shrubs and small trees, the energy savings could heat 57,000 Alaskan homes for a year.[637] In the U.S., Leadership in Energy and Environmental Design ("LEED"), established

---

633. Id.
634. ROGERS & KOSTIGEN, supra note 209 at 138.
635. Oliver, supra note 422.
636. Id.
637. ROGERS & KOSTIGEN, supra note 209 at 138.

by the United States Green Building Council, provides the most commonly used way of determining whether a building is more or less "green" or "sustainable."[638] Green building awards, however, mask the fact that the average green building generates the need to increase automobile use for commuters and those traveling to the building by 137%, suggesting that sustainability ratings should include the reduction of automobile use and the increase in more sustainable forms of transport.[639] LEED awards, offering better buildings with reduced energy use, better passive and sometimes active solar technology, recycled materials, and lowered water use have nevertheless been criticized for making limited improvements and carbon emissions reductions when the technology exists for far more significantly green materials and systems that can provide facades and roofs as solar generators and carbon sinks.[640] Only about 1,000 buildings have been certified since 2000.[641] Despite the fact that some developers cherry-pick low-effort and low-cost LEED point strategies, e.g., installing bamboo flooring and bike racks, and that the cost of certification is relatively high (nearly $100,000), but many developers are using the LEED checklist to build greener buildings without pursuing the costly certification making the program somewhat successful.[642] School construction constitutes the largest sector of non-residential construction projects in the U.S and provides another opportunity for improvement and gains.[643] Green schools that cost but 2% more than traditional schools, but offer broad operational cost savings are better places to learn.[644]

---

638. Jeffrey S. Conner, *Managing the Risks of LEED Certification*, 30 Los Angeles Law. 10 (Jan. 2008); Nancy J. King & Brian J. King, *Creating Incentives for Sustainable Buildings: A Comparative Law Approach Featuring the United States and the European Union*, 23 Va. Envtl. L.J. 397, 406–09 (2005); David G. Mandelbaum, *Corporate Sustainability Strategies*, 26 Temp. J. Sci. Tech. & Envtl. L. 27, 38 (2007); Jonathan Riker, *The Green Zone: Green Building Requirements Must Strike a Balance Between Market Economics and Social Needs*, 30 Los Angeles Law. 27 (Jan. 2008).

639. *Driving to Green Buildings: The Transportation Energy Intensity of Buildings*, Envtl. Building News, Sept. 2007, *available at* BuildingGreen.com, *at* http://www.building-green.com/auth/article.cfm?filename=160901a.xml&printable=yes.

640. Anya Kamenetz, *The Green Standard?: LEED Buiuldings Get Lots of Buzz, But the Point is Getting Lost*, Fast Company, *available at* http//www.fastcompany.com/magazine/119/the-green-standard_Printer_Friendly.html.

641. James S. Russell, *Can LEED Survive the Carbon-Neutral Era?*, Metropolis Magazine, Nov. 2007, *available at* http://www.metropolismag.com/cda/print_friendly.php?artid=3051.

642. *Id.*

643. Eva Steele-Saccio, *Education by Design*, Good Magazine, available at http://www.goodmagazine.com/section/Features/education-by-design.

644. *Id.*

Autofreie Siedlung carfree housing, Munster, Germany.
Source: James A. Kushner.

Unfortunately, 70% of architects are located in northern, developed countries and cities while 70% of the new development is occurring in developing countries where extreme poverty exists and architects are too often focused on aesthetics rather than solutions to needs, sustainability, and liveability.[645]

The United States must reverse the trend toward larger homes that demand more materials, more heating and cooling, more power for lighting and appliances, and often consume more land thereby making densification more difficult and costly. A 3,000 square-foot house over a 50-year lifetime generates 30 to 40 times the weight of the carbon contained in its frame or the equivalent of three 16-miles-per-gallon SUVs driven 12,000 miles per year for more than 50 years.[646] The concrete production and transportation needed for the house generates 47 tons of carbon dioxide and the lumber laid end to end would extend more than four miles.[647] Even if the 3,000 square-foot house utilizes green con-

---

645. Amelia Gentleman, *Architects Aren't Ready for an Urbanized Planet*, INT'L HERALD TRIBUNE, Aug. 20, 2007), *available at* http://www.iht.com/bin/print.php?id+7182262.

646. Stan Cox, *SUVs Without Wheels*, COMMON DREAMS.NEWS CENTER, Mar. 13, 2008, *available at* http://www.commondreams.org/archive/2008/03/13/7650/.

647. *Id.*

struction it consumes 50% more energy than a 1,500 square-foot built only to mediocre energy standards.[648] The design of housing settlements has a profound effect on emissions since design can determine transport choices, e.g., the walkable community and the transit-served community compared to designs that depend on personal automobiles. During the next generation the population of the U.S. will increase by 100 million and jobs will swell by 60 million or a 40% increase in jobs existing at the end of the 20th century.[649] This will require increasing the housing stock by 25% together with another 20% due to disasters and redevelopment;[650] 60% of nonresidential development will be replaced.[651] The new housing will equal one-half of all residential units existing at the end of the 20th century.[652] The combined increase equals roughly 30 billion new square feet, an amount equal to all that existed at the end of the 20th century.[653] Thus, America has the opportunity to rebuild in a more compact, sustainable, and economically efficient, egalitarian design, thereby elevating the very quality of life.[654]

1. Local Response

Local building codes should require green, energy-saving materials and mechanical systems and modestly-angled organic roofs.[655] Even the use of passive solar features, such as skylights or masses of heat-absorbing concrete coupled with sufficient insulation can significantly reduce energy needs.[656] Boston requires that new buildings meet minimum green standards and San Francisco has proposed mandating that all buildings meet an even higher standard by 2012.[657] A Massachusetts-sponsored study found that while green building increased construction costs by 2% that such buildings were on av-

---

648. *Id.*
649. Arthur C. Nelson, Casey J. Dawkins & Thomas W. Sanchez, The Social Impacts of Urban Containment 146 (2007).
650. *Id.*
651. *Id.*
652. *Id.*
653. *Id.*
654. *Id.* at 147–48.
655. Bernow, *in* Climate Change Policy 189, 194–95, *supra* note 93. *See also* Sussman, *supra* note 545 (green building and alternative energy encouragement).
656. Berger, *in* Climate Change Policy 411, 423, *supra* note 428.
657. Robert Selna, *Eco-Tough S.F. Code Proposed for Buildings*, SFGate.com, July 11, 2007.

10 · HOUSING AND CONSTRUCTION

WGL-Terrein car-free housing, Amsterdam, NL.
Source: James A. Kushner.

erage 25 to 30% more energy efficient.[658] San Mateo County, California has proposed a fast track program for building permits that could reduce average waiting times from seven to three weeks when a green building is proposed.[659] Washington, D.C.'s Green Building Act of 2006 includes an incentive program, expedited document review and is funded by a green building fee on all permits.[660] At the turn of the century, 38% of U.S. energy consumption was in buildings and the use of appliances, 36% from industrial applications, and 26% from transportation.[661]

Individual metering rather than a single meter for an apartment complex can reduce consumption as can "green energy programs." Calling for consumers

---

658. *Id.*
659. Rebekah Gordon, *'Green' Builders May Get Fast Track*, PALO ALTO DAILY NEWS, July 25, 2007, *available at* http://www.paloaltodailynews.com/article/2007-7-19-0719-smc-green.
660. D.C. CODE Ch. 14A §§6-1451.01 to 6-1451.11 (Supp. 2007).
661. Berger, *in* CLIMATE CHANGE POLICY 411, 434, *supra* note 428.

to voluntarily pay a power premium targeted toward renewable use can be effective if heavily subsidized.[662] Tax credits should be offered to encourage such participation.[663] Taking preorders for green appliance technology can stimulate and demonstrate market demand and speed up production.[664] Community gardens can make a significant contribution to carbon sequestration, food security, and social capital.[665] Building codes should shift to maximum house size rather than minimum sizes and shift from minimum parking limits to maximum parking limits. Any parking should be a minimum distance and not closer than the closest transit stop. The disability and elderly communities should be accommodated through transit policies. Taxes on developers, housing acquisition, and property should reflect carbon emissions with special rates rewarding conservation. Local zoning codes and master planning should focus on higher density transport corridors; intense development around stops; and mixed use, walkable communities with increased provision for pedestrian trails; dedicated bicycle tracks, parks, and neighborhood gardens and urban agriculture. Building designs should incorporate natural ventilation and materials that insulate to allow the elimination or reduction of air-conditioning.

In 2007, Berkeley, California passed Measure G calling for reducing greenhouse gasses from the city by 80% by 2050.[666] The city is proposing that landlords provide each tenant with a free transit pass while allowing a $7 monthly rent increase; that car sharing cars[667] be located on every block; that all new buildings, resold homes, and renovations replace older appliances with solar-powered water heaters and other high efficiency appliances and upgrade insulation and windows; and that garages be equipped with outlets for electric cars; and mandates green requirements for all buildings, including the use of recycled materials.[668] The city intends to establish assessment districts to fund residential solar panels.[669]

---

662. *Id* at 439.
663. *Id* at 440; King & King, *supra* note 638.
664. Robert M. Margolis & Daniel M. Kammen, *Energy R&D and Innovation: Challenges and Opportunities*, in Climate Change Policy: A Survey 469, 488 (Stephen H. Schneider, Armin Rosencranz & John O. Niles eds. 2002).
665. Sheila R. Foster, The City as an Ecological Space: Social Capital and Urban Land Use, 82 Notre Dame L. Rev. 527 (2006).
666. Jones, *supra* note 315.
667. Beatley, Green Urbanism, *supra* note 170 at 150–56; Kushner, The Post-Automobile City, *supra* note 170 at 99–102; Macht, *supra* note 316; Mannion, *supra* note 316.
668. Jones, *supra* note 315.
669. *Id.*

Zero Energy Houses, The Vauban, Freiburg, Germany.
Source: James A. Kushner.

Chicago, ranked the greenest city in the U.S., has developed 5.2 million square feet of LEED-certified space.[670] Chicago also has developed 2.5 million square feet of green organic roofs on 200 buildings.[671] Sacramento, California requires that all new city buildings meet the LEED silver threshold and is considering an enterprise zone for green technology companies, where taxes would be lowered to encourage such development.[672] Communities could impose a system of impact fees levied on non-green development, using the proceeds to subsidize LEED certified gold and platinum buildings along with exactions measured by the projected use of energy and infrastructure such as water, sewer, and roads.[673]

---

670. Mary Lynne Vellinga, *Aspiring to Be America's Greenest City: State Committment Helps Sacramento Rank No. 2 in Energy Efficient Office Space*, Sacramento Bee, Sept. 5, 2007, available at http:www.sacbee.com/378/v-print/story/360985.html (San Francisco has 1.1 million square feet, Los Angeles, 1.85 million).

671. Emily, *Chicago Green Roof Program*, Inhabitat, Aug. 1, 2006, available at http://www.inhabitat.com/2006/08/01/chicago-green-roof-program.

672. Vellinga, *supra* note 670.

673. Kingsley, *supra* note 626.

2. Regional Response

Regional housing policy can be directed to developing local mass-produced housing and support for industries to develop, install, and maintain ecological systems. Regional planning can also be important to prohibit sprawl and establish planning on a regional level that supports public transport. Regional government regulation, leadership, and cooperation can also better assure that affordable housing is provided in transit-served, mixed-use neighborhoods.

3. State Response

States are in a powerful position to affect housing policy in preempting local laws hostile to policy, imposing planning obligations, and in adjusting tax policy to create incentives for developing sustainable communities. Carbon taxes would encourage emission reductions from motor vehicles, building, and energy generation.[674] Such a carbon tax would be imposed on suppliers of coal, crude oil, and natural gas based upon the carbon content of each.[675] States can support the development of non-carbon-emitting co-generation plants that produce both power and heat.[676] Power companies could be required to phase out production exceeding minimum emission standards and construction of coal-fired plants would be halted.[677]

States could require power providers to annually increase the percentage of power produced from renewables such as wind, solar towers and troughs—large industrialized parabolic solar farms that heat oil which drives a steam turbine capable of generating electricity for 500,000 people, geothermal and hydro sys-

---

674. Berger, *in* CLIMATE CHANGE POLICY 411, 441, *supra* note 428; Burtraw & Portney, *in* NEW APPROACHES ON ENERGY AND THE ENVIRONMENT, *supra* note 549 at 19 (recommending gradual pasing of increases so as not to destroy the value of existing long-lived investments); Goulder & Nadreau, *in* CLIMATE CHANGE POLICY, *supra* note 342 at 120; Harper, *supra* note11 (advocating carbon taxes as superior to regulation or permit trading). *See also* Soares, *in* EU CLIMATE CHANGE POLICY, *supra* note 549 at 256–73; Donehower, *supra* note 116 (arguing that without the incorporation of China and the United States, the world's two largest polluters, the carbon markets may serve as a successful market tool and example of the efficiency of an open market to cost-efficiently solve environmental problems, but will do nothing to curb GHG emissions and limit the effects of climate change).
675. Goulder & Nadreau, *in* CLIMATE CHANGE POLICY, *supra* note 342 at 131.
676. Bernow, *in* CLIMATE CHANGE POLICY 189, 197–98, *supra* note 93.
677. Berger, *in* CLIMATE CHANGE POLICY 411, 441, *supra* note 428.

tems, biomass, and landfill-generated gas.[678] Presently, Texas is requiring an increasing percentage of renewable energy.[679] Currently, The United States derives 92% of energy from fossil fuels and less than 1% from renewables.[680] Nevertheless, wind and solar energy, if increased 8% annually, can replace the need of nonrenewables.[681]

Banks have been reluctant to widely adopt location efficient mortgages which reduce credit and income requirements for those choosing homes that include a significant reduction in transportation cost.[682] The state might require use of the instruments or Congress might require that secondary lenders offer to purchase such mortgages at attractive, above-market prices to create an incentive for lenders. In California, Governor Arnold Schwarzenegger has mandated that all new state office buildings be certified at at least a LEED silver rating and must produce 10% of their energy needs from solar panels or an alternative method.[683] The Cal-EPA state building in Sacramento features waterless urinals and a bin of wriggling worms under each desk to digest lunch scraps.[684]

4. Federal Response

Federal carbon taxes would encourage emission reductions from motor vehicles, building, and energy generation.[685] Such a carbon tax would be imposed on suppliers of coal, crude oil, and natural gas based upon the carbon

---

678. Bernow, *in* CLIMATE CHANGE POLICY 189, 194–95, *supra* note 93. On renewables generally, see Berger, *in* CLIMATE CHANGE POLICY ch 16, *supra* note 428 (contrasting alternatives at 414–18 and discussing minimum contents of portfolio at 436–37 and at 441, recommending increasing portfolio requirements). Twenty-two states set a renewable energy portfolio percentage. *See* PEW CENTER ON GLOBAL CLIMATE CHANGE, LEARNING FROM STATE ACTION ON CLIMATE CHANGE (Mar. 2006), *update available at* http://pewclimate.org/docUploads/UpdatePewStatesBriefMarch2006%2Epdf.
679. Berger, *in* CLIMATE CHANGE POLICY 411, 437, *supra* note 428.
680. *Id* at 411.
681. Loucks, *in* CLIMATE CHANGE POLICY, *supra* note 55 at 502.
682. GOLDSTEIN, *supra* note 314 at 163–64.
683. Vellinga, *supra* note 670.
684. *Id*.
685. Burtraw & Portney, *in* NEW APPROACHES ON ENERGY AND THE ENVIRONMENT, *supra* note 549 at 19 (recommending gradual pasing of increases so as not to destroy the value of existing long-lived investments); Goulder & Nadreau, *in* CLIMATE CHANGE POLICY, *supra* note 342 at 120; Harper, *supra* note 11 (advocating carbon taxes as superior to regulation or permit trading).

content of each.[686] A national, renewable energy bank could provide assistance and offer loans to establish renewable energy.[687] The federal government should dramatically increase research and development appropriations, increase tax subsidies for renewable energy and building applications and provide for technology transfers to developing countries as well as purchase renewable energy[688]

Both states and the federal government could enact tax deductions or credits for retrofitting and installing energy-saving materials and building mechanical systems.[689] Congress should reestablish an affordable housing policy and emphasize rehabilitation over new construction, mandating sustainable building methods and community design by conditioning eligibility on local smart growth planning. Congress might also preempt local zoning and other regulations that impede densification, mixed-use, transit-oriented communities or affordable housing.[690] Foreign aid should target sustainable planning and building construction. Taxing large homes or homes without an energy efficiency retrofit could be tied to tax incentives for retrofit and incentives to generate excess power for renewable sources such as solar, wind, or geothermal.

5. International Response

Internationally, resources must be allocated to those nations unable to invest in sustainable planning and building construction. International agreement should adopt strict carbon caps, carbon taxes, and provide foreign aid to assist in compliance by developing nations and those suffering the worst effects of global climate change.

---

686. Goulder & Nadreau, *in* CLIMATE CHANGE POLICY, *supra* note 342 at 131.
687. Berger, *in* CLIMATE CHANGE POLICY 411, 441, *supra* note 428.
688. *Id* at 442. *See also* Margolis & Kammen, *in* CLIMATE CHANGE POLICY, *supra* note 664 at ch. 18.
689. Bernow, *in* CLIMATE CHANGE POLICY 189, 195–96, *supra* note 93.
690. *Cf.* Peter W. Salsich, Jr., *Toward a Policy of Heterogeneity: Overcoming a Long History of Socioeconomic Segregation in Housing*, 42 WAKE FOREST L. REV. 459 (2007).

Chapter 11

Management of Federal Lands and Agencies

Preserving large segments of our land base as wilderness is essential to maintain a storehouse of natural resources for emergencies and for the future development of medical and agricultural products. Equally important, wilderness provides immediate noncommodity benefits essential to the country's continued vitality. Primitive outdoor recreation and its physical, psychological, and spiritual benefits present a fundamental wilderness value. Moreover, large, undisturbed ecosystems confer invaluable ecological and scientific benefits, such as an essential habitat for numerous species and a baseline database for understanding how healthy ecosystems function. These wilderness values cannot simply be dismissed as some "green" pipedream. They are absolutely essential to maintaining a viable and creative society in the face of current human population pressures.
    Scott W. Hardt, *Federal Land Management in the Twenty-First Century: From Wise Use to Wise Stewardship*, 18 HARV. ENVTL. L. REV. 345, 387 (1994)

The federal government constitutes a vast community of land, buildings, fleets of vehicles, agencies, and military installations around the globe. The federal government owns 655 million acres or 34% of the nearly 2.3 billion acres of land in the United States.[691] If the federal government were to require sustainable transportation, energy production, and conservation, including

---

691. RUTHERFORD PLATT, LAND USE AND SOCIETY: GEOGRAPHY, LAW AND PUBLIC POLICY 6–22, fig. 1-2 (1996) (29% of lands). Carol Hardy Vincent *et al.*, Federal Land Management Agencies: Background on Land and Resources Management, in SAMUEL T. PRESCOTT, FEDERAL LAND MANAGEMENT: CURRENT ISSUES AND BACKGROUND 37, 37 (2003) (34% of lands and buildings). *See generally* FEDERAL LAND MANAGEMENT AGENCIES (Pamela D. Baldwin ed. 2005); SAMUEL T. PRESCOTT, FEDERAL LAND MANAGEMENT: CURRENT ISSUES AND BACKGROUND (2003); PUBLIC LANDS: USE AND MISUSE (William E. Neeley ed. 2007).

Wind Generators on Federal Lands in Palm Springs, California.
Source: U.S. Bureau of Land Management.

the retrofit of government buildings, and condition loans and grants on energy savings and reducing carbon emissions, it would have a significant mitigation impact.

1. Local Response

Local governments in addition to lobbying for federal emissions reductions can, through zoning and building code amendment, assure that future developments and projects meet low emissions standards.

2. Regional Response

On a regional basis, federal facilities can be limited to infill development under a smart growth regime and local governments can be encouraged to adopt sustainable zoning and building code standards together with standards for those agencies contracting with local government.

3. State Response

States can cooperate in pressuring the federal government to go emissions-free or to reduce emissions and can preempt local government on mandating sustainable planning, zoning, development review, and building codes. States also have the power to impose retrofit requirements on federal buildings.

4. Federal Response

The federal government has plenary power over its lands and agencies and can require and finance retrofit and sustainable transportation domestically and abroad. In addition, sustainable buildings and transport can be a condition of federal contracting. Contracts should also be considered on the basis of the lowest emissions, thus rewarding shorter and more sustainable transport components. The federal government should establish solar and wind farms on federal lands. Federal lands can be closed to high-carbon emitting vehicles, including off-road vehicles, snowmobiles, and traditional internal combustion vehicles that do not meet a minimum efficiency standard, such as 40 miles per gallon, and a maximum emissions standard.

5. International Response

International agreements can call for emissions reduction by governments in the management of its agencies, buildings, and lands.

Chapter 12

Oceans and Seas

> Whether wittingly or in ignorance, the deep ocean often serves as a sink for our society's most noxious by-products, from the deliberate dumping of chemical weapons and nuclear wastes to the inadvertent buildup within the deepwater biota of synthetic organic chemicals and mercury.... As our uses of the deep sea continue to grow, so will our need to understand our impacts upon it.
>
> Tony Koslow, The Silent Deep 161 (2007)

Of the 5.4 billion tons (or gigatons) of carbon emissions released into the atmosphere from burning fossil fuels each year, one third is absorbed by the oceans.[692] The gas is extremely soluble and over the centuries, the oceans have absorbed about 80% of carbon emissions or between 1000 to 10,000 gigatons.[693] Unless carbon is transported to the very deep ocean it is metabolized returned to the surface, and released.[694] Carbon dioxide is soluable between a depth of 300 meters, where it first turns liquid, and 3,000 meters but it is buoyant so it rises to the surface.[695] Below 3,000 meters depth, the carbon is denser than sea water and forms a carbon dioxide lake that eventually ices over and dissolves.[696] The solubility of carbon dioxide increases in cold water and decreases in warm water; less carbon dioxide is sequestered as water temperature rises.[697] Oceans are affected by fossil fuel burning and are influencing global climate change by affecting El Niño patterns, increasing acidity through falling particles, and changing and disrupting deep-water currents

---

692. Tony Koslow, The Silent Deep: The Discovery, Ecology and Conservation of the Deep Sea 156–57 (2007).
693. *Id.*
694. *Id* at 157.
695. *Id* at 160.
696. *Id.*
697. *Id* at 157.

that determine climatic conditions.[698] Although the seas are not carbon emitters, carbon has had a profound effect on them. Algae and warming waters threaten coral, plant, fish and mammals. Ocean dumping is a serious problem as is the development of excessively large fish farms. While a body of international law prohibits pollution and dumping, both still occur.[699] Greater enforcement and harsher consequences are required. Overfishing is also a serious concern as it exacerbates the impact on fish and food supply caused by the effects of global warming. The problem is government-caused as extraordinary subsidies are offered to the fishing industry: in 1995, $124 billion were spent to catch $70 billion worth of fish.[700] To restore the fishing resources, at least 20% of ocean area must be placed into marine protection as compared to the current 0.01%.[701] Subsidies should be eliminated and nations should extend their protected waters to 200 miles off shore, prohibit fishing in the zone and support breeding.[702]

Alternative energy also presents a conflict with the health of the seas since alternative energy from off shore wind turbines or tidal and wave power may threaten species and the seabeds.[703] The seas and climate are endangered even if efforts succeed in reducing carbon emissions on the planet:

> [W]hen the temperature of the atmosphere changes, several centuries are required for the depths of the ocean to reach a new thermal equilibrium. For this reason, it has been estimated that even if greenhouse gas emissions dropped back to pre-industrial levels today, sea levels would continue to rise, primarily due to thermal expansion of ocean water, for another 300 years. As bad as some of the near-term changes in climate may be, the greatest damages, arising from today's carbon emissions, will occur far more than one lifetime into the future.[704]

---

698. Gelbspan, Boiling Point, *supra* note 39 at 87.
699. *See generally* Peter Jacques, Globalization and the World Ocean (2006); The Law of the Sea: Progress and Prospects (David Freestone et al. eds 2006); Koslow, *supra* note 692; Alan Longhurst, Ecological Geography of the Sea (2d ed. 2007); Simon Marr, Precautionary Principle in the Law of the Sea: Modern Decision-making in International Law (2003); Waters in Peril (Leah Bendell-Young & Patricia Gallaugher eds. 2001).
700. Martin, The Meaning of the 21st Century, *supra* note 32 at 37.
701. *Id* at 35–36.
702. *Id.*
703. Spalding & de Fontaubert, *supra* note 475.
704. Heinzerling & Ackerman, supra note 54 at 334.

Huge surf with offshore wind North Shore, Oahu, Hawaii.
Source: National Oceanic & Atmospheric Administration.

1. Local Response

Local government can prohibit the serving and sale of endangered fish, farmed fish, and fish transported from distant lands. Global warming and current ocean fishing and breeding threaten sea life and the world's food supply that is diminished through the loss of water and agriculture. The effects of ocean polluting fish farms can be mitigated by prohibiting the sale of foreign-shipped fish. Domestic fish farming should be carefully regulated to assure anti-pollution and healthfulness standards. Local education programs can also instruct on consumer conservation.

2. Regional Response

On a regional basis, communities can be mobilized for conservation efforts and ordinances can be passed on a county-wide basis to prohibit the sale of endangered fish and food that is transported long distances

3. State Response

States are also able to prohibit the importation and sale of endangered species that, along with global warming, threaten food supplies and food products requiring extensive transport. Those states, because of their proximity to the oceans, lakes, or rivers, have a special responsibility for protecting water quality and bear great adverse impacts from declining water quality and declining sea life. Even states that do not border oceans contribute to the problems faced by oceans and have a responsibility to cooperate.

4. Federal Response

The federal government is also able to prohibit the importation and sale of endangered species including fish that may be disappearing due to climate change as well as food products requiring extensive transport. The federal government can also impose high standards on American shipping and fisheries within U.S. waters. A carbon tax could fund research into deep water carbon sequestration, along with storage in saline aquifers or old oil and gas fields.[705]

5. International Response

International agreements can increase standards for harvesting endangered species, the shipment of ocean products beyond a limited distance except in the case of humanitarian aid, and can impose enforceable limits on fishing and dumping as well as standards on shipping, including reduced carbon emissions, safer hulls, and safety equipment. Inessential sea transport including pleasure cruise ships, foreign-shipped goods, oil and natural resources and recyclables should be halted and cleaner technology developed. Oil drilling, sewage and other waste or by-product dumping from mining and other technologies halted and military operations curtailed unless absolutely necessary for defense. Resources should also be shared with poorer nations to permit attainment of international standards.

---

705. KOSLOW, *supra* note 692 at 161.

Chapter 13

Population

> To keep up with the growth in human population, more food will have to be produced worldwide over the next 50 years than has been during the past 10,000 years combined....
>
> Ian Sample, *Global Food Crisis Looms as Climate Change and Population Growth Strip Fertile Land*, GUARDIAN, Aug. 31, 2007

Sustainable development and carbon dioxide emissions reduction is undercut by the rise of world population.[706] The current world population is 6.6 billion and is projected to rise to 9 billion by 2050.[707] As of 2007, 408 cities contain more than one million inhabitants and an additional 2 billion people will reside in cities between 2000 and 2030.[708] As of 2001, 1.3 billion, or one-third of the world population, live in extreme poverty on less than $1 per day.[709] The world population is increasing by 70 million annually placing dramatic strain on energy and food.[710]

The United States reached 300 million population in 2007, all of whom are increasing energy consumption and congesting roads with personal automobiles and the motor vehicles necessary to supply goods and services. Amazingly, the United States is heading for a population of 400 million by the year 2040.[711] The rising demand for energy that accompanies the increasing population is staggering but the added competition for energy with the rising de-

---

706. *See generally* MASSIMO LIVI-BACCI: A CONCISE HISTORY OF WORLD POPULATION (4th ed. 2007).

707. BROWN, *supra* note 36 at 7; MASON, *supra* note 242 at 42; SMITH, *supra* note 108 at 14 (forecasting 8.9 billion by 2050 and leveling off at 11 billion). For the most current population, see http://www.poodwaddle.com/worldclock.swf.

708. Oliver, *supra* note 422.

709. SMITH, *supra* note 108 at 14.

710. Ambrose Evans-Pritchard, *Why the Price of 'Peak Oil' is Famine*, TELEGRAPH.CO.UK, Feb. 9, 2008, *available at* http://www.telegraph.co.uk/money/main.jhtml?xml=/money/2008/02/07/cnoil107.xml.

711. Robert E. Lang & Arthur C. Nelson, *America 2040: The Rise of the Megapolitans*, PLANNING, Jan. 2007.

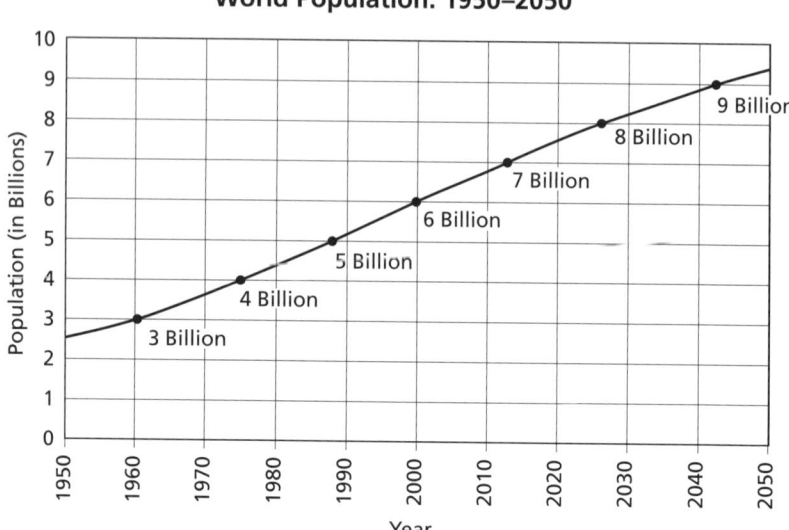

World population projection. Source: U.S. Census Bureau.

mand of developing nations such as China, India, and the population growth in Latin America, Africa, and the rest of Asia is also astounding. Population growth also translates into demand for more food, more deforestation, greater competition for a dwindling water supply and more building. An inverse correlation exists between per capita energy consumption and female fertility.[712] The richer the nation, the higher the rate of energy consumption and the fewer the number of children born.[713] Thus, increasing the wealth of poorer nations is critical to controlling over-population. It is imperative that the wealthy nations assist poorer nations in developing alternative cheaper energy that will emit less carbon as those nations grow their population and infrastructure and use more energy.

The dramatic increase in atmospheric carbon is directly linked to population and calls for strategies to reduce population growth.[714] Population growth is altered either by an increase in the death rate, e.g., through war, emigration,

---

712. GOODSTEIN, *supra* note 498 at 46.
713. *Id.*
714. Meyerson, *in* CLIMATE CHANGE POLICY, *supra* note 15 at ch. 9.

famine, or disease, or by lowering the birth rate.[715] If serious efforts are to be made to save the human race as we know it, climate change must be aggressively attacked. Extending reproductive health services throughout the world will be essential.[716] Ironically, the right-to-life advocates, to support continued life on earth, will be faced with supporting abortion advocacy and abortion services access in the face of the climate change crisis.[717] Although politically unpopular in the United States, family planning, including birth control and abortions must be available to all and free for the poor. Since fertility rates are also inversely related to the level of education, universal education could greatly reduce birth rates.[718] Unfortunately, while family planning services, such as abortion, birth control, vasectomies, and sterilization should be encouraged, subsidized or even mandated, no such programs have ever been successful.[719]

As future work forces dwindle in some nations with declining birth rates, increasing the workforce through higher immigration is a better option than doing so by encouraging births.[720] The lowering of the death rate through public health improvements and medical advances and the increase in life expectancies seriously threatens hopes for the population control that is essential.[721] Several health-related influences may have some impact on lowering overpopulation and carbon generation: a worldwide epidemic of reduction in sperm quality and quantity exists and is accelerating as are testicular deformities and testicular cancer, all of which threaten life in a different way.[722]

---

715. Paul R. Ehrlich, The Population Bomb: Population Control or Race to Oblivion? 34 (1968).

716. Ann Hwang, *The United States Should Support Family Planning Services in Developing Nations*, in Developing Nations 102 (Berna Miller & James D. Torr eds. 2003). *But see* Laura L. Garcia, *The United States Should Not Support Family Planning Services in Developing Nations*, in Developing Nations 124 (Berna Miller & James D. Torr eds. 2003) (questioning efficacy).

717. *Cf.* Meyerson, *in* Climate Change Policy, *supra* note 15 at 256–57 (reporting that a substantial percentage of pregnancies are unwanted).

718. Brown, *supra* note 36 at 124–28; Martin, The Meaning of the 21st Century, *supra* note 32 at 60–62.

719. Ehrlich, *supra* note 715 at 81–94.

720. Judy Dempsey, *Coalition Tackles Germany's Falling Birth Rate: Financial Package for Working Women Wins Wide Support*, Int'l Herald Tribune, Jun. 15, 2006; *Russian Region to Host Day of Conception*, Contra Costa Times, Sept. 11, 2007 (reporting that governor of Ulyanovsk seeking to increase procreation by offering money, cars, refrigerators and other prizes if couples give birth nine months following September 12, the day of conception).

721. Ehrlich, *supra* note 715 at 81–94.

722. Martin, The Meaning of the 21st Century, *supra* note 32 at 137–39.

1. Local Response

Although local government lacks the authority to impact population growth, efforts can be made to encourage family planning and related services to fund services for the poor, and to undertake campaigns to educate the public on birth control, family planning, and the dangers of overpopulation. Improved education for the entire community can also result in a reduced birthrate. Programs to attract and assimilate immigrants should be undertaken in lieu of encouraging births.

2. Regional Response

At the regional or county level, public education campaigns and improved education opportunities together with regional planning for reproductive services would be beneficial. Mandating language and other cultural programs to assist in the assimilation of immigrants should be undertaken.

3. State Response

State government can mandate that local planning include adequate planning for reproductive health service and funding for service to the poor. State tax laws should not reward families with children and should send a clear message of incentives for remaining childless with progressively higher tax rates for families with more than one child. States can also require the extension of education opportunities for the entire community and can fund family planning services and public service educational campaigns to bring awareness.

4. Federal Response

The federal government should pursue a course of funding family planning and reproductive services throughout the United States and the world through foreign aid. Local governments should be obligated to engage in adequate planning and implementation. Tax laws at the federal level can also reward childlessness, abortion, and small families. For example, tax advantages for those with children should be eliminated with tax deductions or credit offered to those who wait to marry until age 30, for not having children, or for undertaking a vasectomy or sterilization. Excise taxes could be imposed on layettes, cribs, diapers, diaper services, expensive toys, daycare, preschool, private schooling, babysitting, and after school care.[723] Government-funded abortions should be available together with birth control and procreation education.[724] Public awareness and public service educational campaigns should be undertaken. The evidence suggests that liberating women to pursue education and professions has a greater effect on lowering the birth rate than even Draconian policies such as China's previous policy of limiting family births by regulation.[725] Although China's policy did reduce births, a pattern of noncompliance existed, particularly by the affluent prompting a new initiative to impose progressive fines based on wealth.[726]

5. International

Population growth is one of the great international problems. International cooperation should include extension of family planning and reproduction services and education to the poorer developing world. Some argue that food should not be exported to nations that fail to undertake population controls.[727] Ironically, food supplies may expand where lands have not been over-farmed and soil destroyed, population decreased and developed nations cease dumping subsidized food that destroys local agriculture markets. The most significant improvements will come from education, particularly for women and employment opportunities for women.

---

723. EHRLICH, *supra* note 715 at 136–38.
724. *Id* at 139–42.
725. MARTIN, THE MEANING OF THE 21ST CENTURY, *supra* note 32 at 62–63.
726. ANI, *Family Planning Rules in China Tightened*, YAHOO NEWS INDIA, Sept. 15, 2007 (reporting that fines will be increased to assure a deterrent effect).
727. EHRLICH, *supra* note 715 at 158–73.

Chapter 14

Smart Growth

> [C]oncentration is the genius of the city, its reason for being. What it needs is not less people, but more, and if this means more density we have no need to feel guilty about it. The ultimate justification for building to higher densities is not that it is more efficient in land costs, but that it can make a better city.
> William H. Whyte, The Last Landscape 339 (2002)

> [T]here is an alternative vision, one that imagines a different future. The future is one in which land is consumed sparingly, landscapes are cherished, and cities and towns are compact and vibrant and green.
> Timothy Beatley & Kristy Manning, The Ecology of Place: Planning for Environment, Economy, and Community 1 (1997)

Global climate change mitigation requires a shift from fossil fuels and, along with conversion from coal and gas-fired electricity generation, transportation must change from automobiles. The suburban low-density urban form must be replaced with a higher density design that can be efficiently served by public transport. Smart growth describes more compact communities that are walkable and connected by public transport. Smart growth reflects an anti-sprawl strategy.[728] One form of smart growth has been the establishment of urban growth boundaries to halt sprawl and protect open space, agriculture, and forest lands.[729] Another variation is the use of urban service districts where infrastructure subsidies and development approval is directed to areas desig-

---

728. Robert H. Freilich, From Sprawl to Smart Growth: Successful Legal, Planning, and Environmental Systems (1999); George Galster et al, *Wrestling Sprawl to the Ground: Defining and Measuring an Elusive Concept*, 12 Hous. Pol'y Debate 681 (2001) (offering alternative definitions based on low value density, continuity, concentration, clustering, centrality, nuclearity, mixed uses, or proximity).

729. Lewyn, *Sprawl, Growth Boundaries and the Rehnquist Court*, supra note 189; Stacey, supra note 189.

nated for urbanization.[730] Really smart growth utilizes both methods but ties and defines urbanization areas to those served by public transport.[731] Portland, Oregon is one example of these principles at work. What is extraordinary about Portland is that half of all development in the past decade has occurred within one block of a streetcar route, raising those property values between 35 and 40%.[732] In light of the carbon emissions crisis, smart growth must be redefined to include sustainable planning and building with local agriculture, local waste management, water conservation, and energy independence through non-fossil-fuel, renewable resources such as solar or wind.

Curiously, outdated zoning schemes across America make it illegal to build classic communities such as Charleston, Savannah, Key West, or Alexandria as well as traditional neighborhoods.[733] Walkable neighborhoods that allow the reduction of automobile use will require the overhaul of American zoning codes. Urban planning for real smart growth[734] and sustainability[735] will require expansion of expertise called for in the context of a legislature, planning commission or board.[736] Greater understanding and sensitivity towards the environment should control regional as well as local planning. Shifting 60% of new growth to compact patterns would avoid 85 million metric tons of carbon dioxide emissions annually by 2030; the savings would equate to a 28% increase in federal vehicle efficiency standards to 32 miles per gallon by 2020.[737] Reducing carbon emissions is linked to establishing compact high-density settlements, eliminating high carbon emitting internal combustion-fueled vehi-

---

730. Gerrit-Jan Knaap, & John W. Frece, *Smart Growth in Maryland: Looking Forward and Looking Back*, 43 IDAHO L. REV. 445 (2007). *But see* Timothy B. Wheeler, *Searching for Signs of Intelligent Growth*, HARTFORD COURANT, Sept. 2, 2007 (finding little benefit of smart growth in Maryland after ten years; while 75% of homes built within growth areas, 75% of land on which homes built outside of designated growth areas).

731. James A. Kushner, *Smart Growth, New Urbanism, and Diversity: Progressive Planning Movements in America and Their Impact on Poor and Minority Ethnic Populations*, 21 UCLA J. ENVTL L. & POL'Y 45 (2002/2003).

732. *Driving to Green Buildings, supra* note 639.

733. Chad D. Emerson, *Making Main Street Legal Again: The SmartCode Solution to Sprawl*, 71 MO. L. REV. 637, 637 (2006).

734. Kushner, *Smart Growth, New Urbanism, and Diversity, supra* note 731.

735. *See generally* John C. Dernbach, *Sustainable Development as a Framework for National Governance*, 49 CASE W. RESERVE L. REV. 1, 3 (1998). See also Timothy Beatley & Richard Collins, *Smart Growth and Beyond: Transitioning to a Sustainable Society*, 19 VA. ENVTL. L.J. 287, 297–99 (2000).

736. *See generally* BULKELEY & BEWTSILL, *supra* note 313.

737. EWING ET AL., *supra note* 632 at 9.

cles, and converting to renewable energy. Changes in transportation and how it integrates with jobs, housing, commercial, and recreation must drive the process to design the growing region. Current city design is one for automobiles rather than people.[738]

If just one in ten new home buyers selected a home site that cut the length of the work commute by one mile each way, 2.4 million gallons of gas could be saved each year, enough to fuel more than 10,000 road trips between Los Angeles and New York City, substantially reducing carbon emissions.[739] Studies have shown that New Yorkers generate 7.1 metric tons of greenhouse gases annually compared to 24.5 generated by the average American and that miles driven dropped between 32 and 43% in Chicago, Los Angeles, and San Francisco as the density of neighborhoods doubled.[740] The larger and denser a city, the more sustainable its form and fewer carbon emissions.[741] With walkable more compact development, people drive 20 to 40% less and enjoy lower costs and improved health.[742] Unfortunately, Americans are driving more as urban sprawl expands. Between 1950 and 1994, the number of miles driven annually in the U.S. increased by 140% while the population increased by only 50%.[743] A study of Seattle, Washington found that residents of walkable communities drove 26 fewer miles daily than those residing in sprawling

---

738. JAMES A. KUSHNER, HEALTHY CITIES—THE INTERSECTION OF URBAN PLANNING, LAW, AND HEALTH (2007); KUSHNER, THE POST-AUTOMOBILE CITY, *supra* note 170 at 76–84; Michael Lewyn, *How Government Regulation Forces Americans into their Cars: A Case Study*, 16 WIDENER L.J. 839 (2007).

739. ROGERS & KOSTIGEN, *supra* note 209 at 136.

740. John Norquist, *We Would Use Less Energy Living Closer Together: Cities Have Powerful Environmental Advantages: They Make it Easier to Walk and Use Public Transit*, PHILLY.COM, May 17, 2007, *available at* http:www.philly.com/inquirer/opinion/20070517_We_would_use_less_energy_living_closer_together.html. *See also* Kate Sheppard, *The Urban Revival: Cities May be the Key to Curbing Climate Crisis*, MSN, Aug. 30, 2007, available at http://stopglobalwarming.msn.com/article.aspx?cp-documentid=5288633 (the average house in Chicago generates 2.5 tons of carbon dioxide each year while the average home in the exurbs generates 11.5 and suburban homes with commuter rail access generate 9; while residents of Manhattan use 25% the energy consumed by the average American, thus sprawl uses 40 times as much land and 4 times the energy).

741. Lance Hosey, *Is Bigger Better?: New Research Supports Eco-Friendly Urbanism*, ARCHITECT MAGAZINE, Aug. 1, 2007 (reporting study by Geoffrey West of the Santa Fe Institute).

742. EWING ET AL., *supra note* 632 at 4.

743. KAHN, *supra* note 25 at 113.

744. EWING ET AL., *supra note* 632 at 6. *See also* KAHN, *supra* note 25 at 113 (31% more miles driven by suburban households over those in center cities).

areas.[744] A meta-analysis of the collected studies on compact development disclose that households residing in developments at twice the density, diversity of uses, accessible destinations, and interconnected streets when compared to low-density sprawl areas drive about 33% less.[745] The average Atlanta houshold would drive 25% fewer miles if relocated to relatively compact Boston.[746]

A review of 23 studies comparing low density and compact communities found that residents of compact communities averaged 8% fewer total miles driven with a maximum reduction of 31.7%.[747] Utilizing transportation models and the EPA's Smart Growth Index, research disclosed that infill locations would generate 35% less driving and emissions than traditional, low-density sites.[748] Gasoline consumption would be reduced by 20 to 30% in sprawling cities like Houston and Phoenix if their urban structure more closely resembled Boston or Washington, D.C.[749] The same family that consumes 1,407 gallons of gas annually in Houston, would use only 919 gallons if it resided in San Francisco.[750] An additional cause of sprawl and vehicle miles traveled is wealth accumulation; with each 10% rise in household income, annual household vehicle miles driven increases by 14%.[751] In addition, 19.1% of suburban households own sport utility vehicles that get lower miles per gallon as compared to 12.1% of urban households.[752] Sprawl also reduces walking, cycling, and public transport use. In 1960, 22 percent of workers took public transport or walked and 64% drove; by 1980, 6.4% took public transport, 5.6% walked, and 84% drove.[753] Public transport use declined to 5.3% in 1990 and was down to 2.7% by 2000.[754] In old transit cities public transport use fell from 30% in 1970 to 23% in 1990; but in bus-only cities, transit users fell from 5% to 2%.[755]

Transport service typically falls as communities sprawl. While car travel takes two minutes per mile for trips under five miles and the bus takes three minutes

---

745. EWING ET AL., *supra note* 632 at 6. *See also* KAHN, *supra* note 25 at 113 (31% more miles driven by suburban households over those in center cities).
746. KAHN, *supra* note 25 at 113.
747. EWING ET AL., *supra note* 632 at Executive Summary.
748. *Id.*
749. KAHN, *supra* note 25 at 114.
750. *Id.*
751. *Id* at 113.
752. *Id* at 116.
753. *Id.*
754. *Id.*
755. *Id* at 117–18.

per mile, the average bus commuter waits 19 minutes for the bus to arrive.[756] In 2000, 10.2% of workers living within five miles of the central business district used public transport compared to 5% of those residing more than five miles away.[757] In Boston, according to the 1990 census, 36% of workers living and employed in the city used public transport compared to 5% of those living and working in the suburbs.[758] The decline in public transport use also results in reduced public support for transport finding.[759] Although the cost of gasoline and automobile transportation along with traffic congestion and awareness of global climate change are increasing popular support for public transport,[760] in the 14 cities that have invested in rail systems over the past thirty years, the share of commuters taking public transport has actually dropped from 8% in 1970 to 6% in 2000. Rising suburban employment and rising intersuburban commuting makes rails to the downtown less efficient and finds few drivers abandoning cars and bus riders may switch to automobiles resulting in lower bus use.[761] Nevertheless, between 1994 and 2004, Boston and Washington, D.C. increased public transport ridership by 27 percent,[762] suggesting that with an expanding demand for urban living, convenient transport, and densification around transit stops a turnaround period for public transport has begun.

Sprawl and suburbanization has also increased the urban footprint as suburban homeowners consume more space. Suburban homeowners have 35% more exterior space and 6% more interior space than central city homeowners.[763] Between 1982 and 1992, the U.S. lost 4 million acres of farmland to suburbanization, an area equal to two-thirds the size of Vermont.[764] Increasing residential density from four units per acre to five units per acre would reduce water consumption by 10%.[765]

---

756. *Id* at 118.
757. *Id* at 118–19.
758. *Id* at 118.
759. *Id* at 119–20.
760. Transportation Finance at the Ballot Box: Voters Support Increased Investment & Choice (2007) (over the last five years communities in 33 states have approved more than $70 billion in new transportation investment with an average success rate of ballot measures of 70 percent—130 measures of 202), *available at* http://www.cfte.org/success/TrendsReport00-05.asp.
761. KAHN, *supra* note 25 at 121.
762. *Id.*
763. *Id* at 121–22.
764. *Id* at 122.
765. *Id* at 124–25.

French Quarter, New Urbanist village, Tübingen, Germany.
Source: James A. Kushner.

The American dream has shifted from the detached home in the suburbs with 55% indicating a preference for living in a smart growth community.[766] Not only has the American dream shifted to smart growth, but a recent survey conducted by the National Association of Realtors and Smart Growth America disclosed that 75% of Americans believe that improving public transportation is a better long-term solution for reducing traffic congestion than building new roads;[767] 26% of Americans believe developing communities that reduce the need to drive would be a better alternative.[768] The same survey disclosed that the top three concerns of Americans regarding growth and development are loss of farmland to development (72%), increased traffic congestion and commute times (70%), and loss of open land such as fields and forests (70%).[769] The

---

766. EWING ET AL., *supra note* 632 at §1.5 (a third of the market would choose smart growth if there was a comparable commute to conventional neighborhoods, more than another quarter if the location was closer to employment thereby reducing commuting time).

767. National Ass'n of Realtors, News Release, *Americans Prefer to Spend More on Mass Transit and Highway*://www.realtor.org/press_ro *Maintenance, Less on New Roads*, Oct. 24, 2007, *available at* http om/news_releases/2007/nar_smart_growth_survey_2007.html.

768. *Id.*

769. *Id.*

immigrants to the United States will be searching for affordable homes in walkable neighborhoods with access by public transport to jobs. Twenty-one percent of Americans over the age of 65 do not drive and the aging population will increasingly require compact walkable communities with public transport or extensive programs of van service.[770]

1. Local Response

Smart growth begins with modifying the community's comprehensive plan and inviting the public to participate in a planning process that can look to all the elements needed for local survival. Coordination on a regional level will be critical on matters of regional concern such as transport, water, and locations for controversial land uses. High density, mixed-use, walkable communities connected by transit must form the foundation. Energy use should be focused on renewable sources and sustainable green building codes. Zoning should reflect a reduction in automobiles and parking, the inclusion of green architecture, organic roof gardens, an increase in tree planting, and the expansion of urban gardening and agriculture. Changing zoning, parking, and subdivision regulations in just the 50 largest American metropolitan areas would quickly bring more housing, neighborhood, and transportation choices to about 168 million people, or more than half of all Americans.[771]

Local government must engage in a new master plan and work with neighboring regional partners to assure emission reductions in public services and transportation between neighborhoods, shopping, and employment centers. Local government will need to study the placement of stations, projected growth, and alternative infrastructure strategies. New public transport systems and lines require densification within 1 to 2 miles of stations. More parks and gardens should be part of local and urban agriculture. An expanded and a redesigned education system is needed to emphasize new technologies, and employee-seeking, high technology corporations should be a part of urban policy. New building codes should integrate best practices in reducing emissions and energy together with more efficient use of water and sanitation, offering carbon sequestration, such as through roof gardens and organic roofs, and the inclusion of solar, wind, or other non-fossil fuel-based

---

770. Haya El Nasser, *Senior Transportation a Growing Concern*, USA TODAY, Dec. 2, 2007, *available at* http://www.usatoday.com/news/nation/2007-12-02-transport_N.htm.
771. EWING ET AL., *supra note* 632 at §7.1.8.

power generators. Densification should mean high density housing over shops, schools, libraries, and offices. Decisions should be made to develop regional and local production of goods and food to reduce transportation-generated emissions.

Although many cities are undertaking the development of new transit lines and embarking on transit-oriented development around stops, the housing developments are occupied by automobile users and very few transit patrons. The development of housing is actually dispersing employment locations away from transit-served areas.[772] For example, at a transit-oriented project in Long Beach, California on the Blue Line trolley to Los Angeles, 6.3% of occupants use the line to commute while 78% have never used the line.[773] Employment, recreation, shopping, and housing destinations must be linked by significantly expanded systems to capture riders away from their cars. Rents should be lowered for those using transit, parking significantly reduced, and a limited number of parking spaces located away from the building. Fares should be eliminated or transit passes provided to those who are carless and relying on transit.

2. Regional Response

At the regional level, the coordination of public transport and rail infrastructure with regional governments is essential. In addition, cooperation should exist in siting controversial uses and sharing regional tax base[774] to encourage community-wide smart growth policies. Regional response should be to conform county plans to the standards of local government and work to bring the entire region into agreement on tax revenue sharing,[775] planning for public transport, open space protection, and integrated densified communities around stops and along transit corridors.[776] Further, planning for all uses needed in the region should be coordinated.

---

772. Sharon Bernstein & Francisco Vara-Orta, *Near the Rails but Still on the Road*, L.A. TIMES, Jun. 30, 2007.

773. *Id.*

774. ORFIELD, *supra* note 168 at 105–108; RUSK, *supra* note 168; Note, *Making Mixed-Income Communities Possible*, *supra*, note 168.

775. ORFIELD, *supra* note 168 at 105–108; RUSK, *supra* note 168; Note, *Making Mixed-Income Communities Possible*, *supra*, note 168.

776. ORFIELD, *supra* note 168 at 105–108; RUSK, *supra* note 168; Note, *Making Mixed-Income Communities Possible*, *supra*, note 168.

3. State Response

State subsidies for transport, housing, and infrastructure should be awarded based on smart growth planning compliance. Tax policy should reward those participating in infill and densification and programs for retrofitting homes and businesses with energy reduction features. In 2006, California enacted A.B. 32, known as the Global Warming Solutions Act of 2006.[777] The law directs the Air Resources Board, the state agency charged with monitoring and regulating sources of greenhouse gases that cause global warming, to reduce emissions.[778] That state's Attorney General filed a lawsuit against San Bernardino County alleging a violation of the statute in approving a new general plan that failed to account for increases in greenhouse gas emissions and proposed to approve new low density housing developments and facilities that would generate traffic.[779] The lawsuit sought mitigation components and, the Attorney General pledged to file similar lawsuits against other non-complying counties.[780] The state's attorney general, former Governor Jerry Brown, is interpreting the law, which seeks to roll back greenhouse gas emissions by 20% by 2025, and requiring offsets for new projects that will generate new emissions.[781] The law itself does not specifically impose such a duty on local government but neither does it preempt efforts of other agencies and other legal mandates.[782] Thus, it may be reasonable that the law recognizes the extraordinary

---

777. Stats. 2006, c. 488 (A.B. 32), *codified at* CAL. HEALTH & SAFETY CODE §§ 38500–38574 (West Supp. 2007). *See* Sajadi, Keeana, Comment, *The Terminator a Trendsetter? How California's Global Warming Solutions Act Will Impact California, the United States, and the World*, 21 J. NAT. RESOURCES & ENVTL. L. 143 (2006–2007); Nordberg, *supra* note 166.

778. CAL. HEALTH & SAFETY CODE § 38510 (West Supp. 2007). See also Patricia Weisselberg, Comment, *Shaping the Energy Future in the American West: Can California Curb Greenhouse Gas Emissions from Out-of-State, Coal-Fired Power Plants Without Violating the Dormant Commerce Clause?*, 42 U.S.F.L. REV. 185 (2007) (arguing that the Greenhouse Gas Emissions Performance Standard Act which also forbids California utilities from making long-term financial investments or procurement contracts with power plants whose GHG emissions exceed the performance standard does not violate the Commerce Clause).

779. Mark Martin, *Sprawl Clashes with Warming in California*, SAN FRANCISCO CHRONICLE, May 27, 2007; John Ritter, *Calif. Sees Sprawl as Warming Culprit*, USA TODAY, Jun. 14, 2007.

780. Martin, *supra* note 779; Ritter, *supra* note 779.

781. Samantha Young, *Ca. Land-Use Dispute Complicates Budget*, FORBES.COM, July 26, 2007, available at http://www.forbes.com/feeds/ap/2007/07/26/ap3956150.html (reporting partisan bickering over the law, the budget and the pending litigation).

782. CAL. HEALTH & SAFETY CODE § 38574 (West Supp. 2007).

danger from greenhouse emissions and require that they be analyzed in environmental impact reports on discretionary land use and other projects, and should ideally and consistently be considered when creating or amending a general plan. The lawsuit has been settled by an agreement of the county to add a reduction of greenhouse emissions policy to its general plan within 30 months; to prepare inventories of 1990 and current emissions data; to project emissions through 2020 and establish a reduction goal and mitigation measures, including replacing all the county-owned diesel vehicles, at an estimated cost of $500,000.[783]

The California Attorney General's office has also notified San Diego that its 193 designated smart growth areas were too numerous and too widely dispersed and should be re-targeted to commuter-train, transit, or fast-bus stops and corridors.[784] The attorney general's active review of local planning could result in improved planning without the necessity of resorting to litigation and brings the state level scrutiny of local planning not seen in California since Jerry Brown served as governor.

Although in this book I have avoided reducing climate change to a litigation strategy and focusing on courtroom oriented solutions, litigation can be an effective spur to regulatory and institutional change so that statutory or common law nuisance doctrine should be employed to halt policies and facilities that generate significant carbon dioxide emissions.[785] One problem with using

---

783. Matt Wrye, *State Drops its Lawsuit*, SAN BERNARDINO COUNTY SUN, Aug. 22, 2007.

784. Dave Downey, *Smart Growth Policy Revision Needed, Some Say*, NORTH COUNTY TIMES, Dec. 6, 2007, *available at* http://www.nctimes.com/articles/2007/12/04/news/top_stories/1_04_1512_3_07.txt.

785. Matthew D. Adler, *Corrective Justice and Liability for Global Warming*, 155 U. PA. L. REV. 1859 (2007); Kirsten H. Engel, *Harmonizing Regulatory and Litigation Approaches to Climate Change Mitigation: Incorporating Tradable Emissions Offsets Into Common Law Remedies*, 155 U. PA. L. REV. 1563 (2007) (advocating Tradable Emissions Offsets as remedial elements together with damages); Farber, *Basic Compensation, supra* note 88 (recommending relief in the form of monitoring, protecting, restoring, or providing substitutes for existing resources over damages); Kevin T. Haroff & Katherine Kirwan Moore, *Global Climate Change and the National Environmental Policy Act*, 42 U.S.F.L. REV. 155 (2007) (arguing that NEPA is not likely to present a viable strategy to challenge global warming and recommending statutory initiatives); Hari M. Osofsky, *Local Approaches to Transnational Corporate Responsibility: Mapping the Role of Subnational Climate Change Litigation*, 20 GLOBAL BUS. & DEV. L. J. 143 (2007); Joni Hersch & W. Kip Viscusi, *Allocating Responsibility for the Failure of Global Warming Policies*, 155 U. PA. L. REV. 1657 (2007) (noting that emissions cases are difficult as they cannot be traced to particular inadequacies on behalf of likely defendants but recommending a pollution tax as a fair and effective remedy); David Hunter & James Salzman, *Negligence in the Air: The Duty of Care in Climate Change Litigation*, 155

tort recovery strategies to deal with climate change is that the damages are unknown and incalculable.[786] An alternative strategy would be to argue through litigation that activities leading to urban sprawl or global climate change constitute state law nuisances allowing legislative regulation, injunctive relief or damage recovery.[787] The California legislature is considering other pending bills that would impose specific obligations to reduce carbon emissions on local government.[788]

Connecticut requires every community to include a planning element in its comprehensive plan providing for integrating transportation and land use so as to establish pedestrian-friendly transit-accessible communities with high density housing and mixed-use development around transit stops.[789] Massachusetts has established a voluntary program that rewards municipalities that establish zoning districts that encourage affordable housing near public transportation and existing city and town centers or other desirable areas.[790] Cities and towns can qualify for up to $600,000 for adopting a district and $3,000 for each building permit issued.[791]

States are in a powerful position to advance sustainable planning and development. Smart growth policies would include mandating a statewide plan for open space, forestation, food supply and most importantly, transportation by public transport and rail development. Planning laws can mandate local and regional governments undertaking sustainable master plans that call for dense, mixed-use neighborhoods around transit stops and including a polycentric mix of suburban communities served by transit. Siting of retail, industry, and office centers would be limited to transit-served land. Brownfield clean-up and infill should also be a priority as would investment in alternative energy. Tax policy and land use control regulation should encourage higher density tran-

---

U. Pa. L. Rev. 1740 (2007); Eric A. Posner, *Climate Change and International Human Rights Litigation: A Critical Appraisal*, 155 U. Pa. L. Rev. 1925 (2007).

786. Heinzerling & Ackerman, supra note 54.

787. Christine A. Klein, *The New Nuisance: An Antidote to Wetland Loss, Sprawl, and Global Warming*, 48 B.C.L. Rev. 1155 (2007).

788. *SB 375 Connects Land Use and AB 32 Implementation*, Planning Report, July, 2007), *available at* http://www.planningreport.com/tpr/?module=displaystory_id=1257&format=html. *See also* Cal. Health & Safety Code §§ 38501–38599 (West 2006) (AB 32).

789. Conn. Gen. Stat. Ann. § 16a-27 (West 2005).

790. Mass. Gen. Laws Ann. Ch. 40R §§ 1–14 (West. Supp. 2007); Robert Preer, *Housing Deal Gets Popular: Towns Just Hope State Ponies Up Aid*, Boston Globe, Sept. 2, 2007, *available at* http://www.boston.com/news/local/articles/2007/09/02/housing_deal_gets_popular?mode=PF.

791. Mass. Gen. Laws Ann. Ch. 40R §§ 1–14 (West. Supp. 2007); Preer, *supra* note 790.

sit-oriented residential, retail employment centers, recreation, mixed use development. Below-market loans and grants should provide incentives for moves to transit-oriented communities, central cities, and brownfields.

4. Federal Response

The federal government can adjust the tax system to provide deductions and credits for smart growth activities and can condition infrastructure, housing, and transportation subsidies on smart growth planning and development. Foreign aid should also be targeted to encourage smart growth for developing cities and nations. The primary influence of the federal government should come in (1) investment in public transport, high-speed rail, and intercity rail, (2) reestablishing a housing subsidy system that will empower local communities to design economically integrated transit-oriented communities, (3) investment in research and development of energy, transport, housing, and food. Federal infrastructure and planning grants should be conditioned on local smart growth planning and implementation. Tax laws can establish incentives for alternative energy production, green architecture, and sustainable development. Foreign aid should emphasize sustainable development and public transport integration with land use and sprawl control. Federal law should preempt state and local unsustainable policies.

5. International Response

International cooperation should reallocate resources to assist the poorer nations to establish smart growth urban policies. International organizations through treaties and protocols should establish planning norms and allocate funds to assist poorer nations in implementation.

Chapter 15

Species Protection

> In a time of war, truth is the first casualty. In a time of warming, it begins with ice. That is followed in short order by plants, animals, fish, birds, and entire ecosystems that are right now migrating toward the poles all over the world in the futile pursuit of stable temperatures. The next set of casualties is people. Only the bugs love warming.
>
> Ross Gelbspan, Boiling Point 19 (2004)

Species protection is essential for the loss of species can have unanticipated adverse consequences.[792] The current mass extinctions of flora and fauna, while not the first to occur, is perhaps the first that has been caused by human behavior.[793] By 2100, 1.25 million species will be extinct should Earth's temperature rise 2.5 to 10.4° F as projected.[794] By 2050, two-thirds of the world's current 16,000 polar bears will be killed from melting Arctic ice according to the U.S. Geological Survey within the Interior Department.[795] The best chance for polar bear survival would be the development of a recovery plan under the Endangered Species Act.[796] The Arctic sea ice has already melted from 5.05 mil-

---

792. *See generally* George C. McGavin, Endangered: Wildlife on the Brink of Extinction (2006); Wayne Hsiung & Cass R. Sunstein, *Climate Change and Animals*, 155 U. Pa. L. Rev. 1695 (2007); Jason Scott Johnston, *Desperately Seeking Numbers: Global Warming, Species Loss, and the Use and Abuse of Quantification in Climate Change Policy Analysis*, 155 U. Pa. L. Rev. 1901 (2007).

793. 3 Johansen, *supra* note 1 at 533.

794. *Id* at 533–34.

795. Associated Press, *Most Polar Bears Could Die Out by 2050: U.S. Geological Survey Says Two-Thirds Could Vanish Because of Ebbing Ice*, Msnbc, Sept. 8, 2007, *available at* http://www.msnbc.msn.com/id/20645362/.

796. J.B. Ruhl, *Climate Change and the Endangered Species Act: Building Bridges to the No-Analog Future*, 88 B.U. L. Rev. 1 (2008) (arguing that the ESA should not be used to regulate greenhouse gas emissions, but rather that it should be focused on establishing protective measures for species that have a chance of surviving the climate change transition and establishing a viable population in the future climate regime and that the ESA can help ensure that human adaptation to climate change does not prevent other species

Polar bear. Source: NOAA Climate Program Office NABOS 2006 Expedition/Mike Dunn.

lion square miles in 2005 to 4.75 million square miles in 2007, a loss of 250,000 square miles.[797] As predators are lost, certain species may thrive and bring about losses of agriculture, plants, or trees, and may expand the possibilities for disease and other changes. The loss of fish can affect food supply and the quality of oceans.

The plight of salmon is an example of poor husbandry. The Bush Administration wanted to develop a $5 billion fish farming industry. Unfortunately, such a farming operation generates as much nitrogen discharge as untreated sewage from 17 million people.[798] Salmon farms in Scotland discharge the amount of waste generated by 9 million people, twice the population of that nation.[799] Four tons of inexpensive fish must be fed to the salmon to produce

---

from adapting as well); Laura Navarro, Comment, *What About the Polar Bears? The Future of the Polar Bears as Predicted by a Survey of Success under the Endangered Species Act*, 19 VILL. ENVTL. L.J. 169 (2008) (although the future is unpredictable, a recovery plan holds some hope).

797. Associated Press, *supra* note 795.
798. SINGER & MASON, *supra* note 98 at 124.
799. *Id* at 123.

one ton of farm salmon. The inexpensive fish would have been the food for cod and haddock and could have fed many coastal people in developing nations instead of exacerbating global warming-caused food shortages.[800] A half a million farm fish escape to the wilds each year often carrying parasites and disease to the wild fish they breed with and infecting them with high levels of sea lice. Farmed salmon have been found to contain artificial coloring fed to them to give them the pink hue of a wild salmon normally achieved from ingesting krill.[801] This practice promotes consumer fraud by allowing farmed salmon to be sold as wild.[802] Shrimping catches, particularly in Asia and Latin America, include as much as one-third endangered sea life such as turtles. Bottom trawling, the common form of shrimping, can destroy coral formations and mangrove forests are often destroyed to make way for shrimp farms.[803] Shrimp farms raise 120,000 tons of shrimp annually but cause an annual loss of 800,000 tons of harvestable wild fish.[804] Only the United States and Canada engage in responsible shrimping.[805]

1. Local Response

Species plans should be undertaken and those endangered be protected, e.g., such as through species migration corridor protection, fishing limits, and the expansion of wetlands and forests. Activities that threaten animal and plant life, including the consumption or sale of endangered species, should be prohibited.

2. Regional Response

Corridors for species migration are required to assist in avoiding extinction.[806] These corridors are best planned for on a regional basis. On a regional

---

800. *Id.*
801. *Id* at 122–24.
802. *Id. See also In re* Farm Raised Salmon Cases, 175 P.3d 1170 (Cal. 2008) (Federal Food, Drug, and Cosmetic Act does not preempt California consumer suit against food chains that allegedly failed to comply with labeling requirements regarding color additives used to make farm-raised salmon look like their "wild" counterparts).
803. SINGER & MASON, *supra* note 98 at 125–28.
804. MARTIN, THE MEANING OF THE 21ST CENTURY, *supra* note 32 at 75.
805. SINGER & MASON, *supra* note 98 at 128.
806. Turman, *in* CLIMATE CHANGE POLICY, *supra* note 44 at 106, *citing* 2 CALIFORNIA ENERGY COMMISSION, GLOBAL CLIMATE CHANGE POTENTIAL IMPACTS & POLICY RECOMMENDATIONS (1991); C.B. FIELD *ET AL.*, CONFRONTING CLIMATE CHANGE IN CALIFORNIA:

level other activities can be planned such as the expansion of wetlands and forests.

3. State Response

States can provide expertise in assisting local communities to undertake species protection plans and can regulate to protect endangered species.

4. Federal Response

The Endangered Species Act[807] could be expanded to protect species in all locations and to monitor imports of food and goods. The principle should also be applied to foreign aid and foreign development activities.

5. International Response

International cooperation should involve protection of species and reallocating resources to assist poorer nations in protecting species.

ECOLOGICAL IMPACTS ON THE GOLDEN STATE (1999); J.B. KNOX & A. FOLEY, GLOBAL CLIMATE CHANGE AND CALIFORNIA: POTENTIAL IMPACTS AND RESPONSES (1992); J.B. SMITH & D.A. TIRPAK, THE POTENTIAL EFFECTS OF GLOBAL CLIMATE CHANGE ON THE UNITED STATES (1990).

807. Endangered Species Act of 1973, 16 U.S.C. §§ 1531–44 (2000).

Chapter 16

Technology

> Beyond avoiding responsibility, both the Clinton and Bush administrations have argued that the U.S. is actually saving the world by investing billions in developing new, low carbon technologies. It is true that many of the most exciting developments have come from the United States. But tackling climate change, like dieting, is as much about what you don't do as what you do. Developing low carbon technologies without cutting your emissions is like eating two Big Macs, four donuts, and an ice cream sundae and then, to be healthy, also eating a salad. Unless new technologies replace fossil fuel burning—rather than simply supplementing it—they cannot reduce a nation's emissions.
>
> George Monbiot, Heat: How to Stop the Planet from Burning vi (2007)

Technology is central to the survival of the planet's people. Research and development resources must look to new systems for generating renewable energy, providing the best transport, and establishing the best systems for water and waste management. There exist a myriad of social, economic barriers to the transfer of technologies to developing nations including the transfer of environmentally sound technologies.[808]

1. Local Response

Local government can expand technical and scientific education in its local schools, community colleges, and universities and establish cooperative ventures between business and educational programs for apprenticeships. Retired

---

808. James Shepherd, *The Future of Technology Transfer Under Multilateral Agreements*, 37 Envtl L. Rptr. News & Analysis 10547 (2007).

University of Minnesota Solar car passes through Lake Benton, Minnesota the "original wind power capital of the Midwest" during the 2005 North America Solar Challenge. Source: U.S. Department of Energy.

engineers, mathematicians, and scientists could participate as part of the infrastructure to improve local education resources. Local government can also assemble land and establish incentives to develop technology incubation parks. Such parks would aid in making communities more compact and less sprawling so that automobile use can be reduced and public transport expanded. Subsidies and other incentives should be provided to encourage alternative energy, green technologies, and solar-based industries.

2. Regional Response

On the regional level, coordination of local governments and leadership in developing the educational infrastructure necessary to allow technology development is required. Each community might have a different focus, such as environmental science, energy development, transportation, mathematics, etc.

3. State Response

The state can influence universities and colleges to expand needed technology curriculum through funding and can direct local schools to cooperate and expand science and technology classes that can be made available to the young and older workers in need of retraining.

4. Federal Response

The federal government through spending can fund engineering, science, and other technology programs in colleges, universities, and high schools and can create incentives for students seeking loans and grants to pursue those fields needed for the future technological economy. Programs of grants and tax incentives such as tax credits, deductions, and accelerated depreciation for developing carbon-free technologies and research funding should be included in the federal strategy.[809] Heat- and drought-resistant crops need to be developed.[810] Foreign aid should emphasize technology transfer.

5. International Response

International cooperation should focus on technology transfer and the reallocation of resources toward the poorer nations.

---

809. Goulder & Nadreau, *in* CLIMATE CHANGE POLICY, *supra* note 342 at 120; Stephen H. Schneider & Kristin Kuntz-Duriseti, *Uncertainty and Climate Change Policy*, *in* CLIMATE CHANGE POLICY: A SURVEY 53, 78 (Stephen H. Schneider, Armin Rosencranz & John O. Niles eds. 2002).

810. Turman, *in* CLIMATE CHANGE POLICY, *supra* note 44 at 106, *citing* 2 CALIFORNIA ENERGY COMMISSION, *supra* note 806; FIELD ET AL., *supra* note 806; KNOX & FOLEY, *supra* note 806; SMITH & TIRPAK, *supra* note 806.

Chapter 17

Transportation

America finds itself nearing the end of the cheap-oil age having invested its national wealth in a living arrangement—suburban sprawl—that has no future.

James Howard Kunstler,
The Long Emergency 17 (2005)

Al Gore outlined a number of personal strategies to reduce energy use in transportation, including (1) emission reduction in vehicles and transit, (2) walk, bike, and take transit while reducing car miles driven, (3) drive smart by off-peak hour driving at slower speeds, (4) purchase a more efficient vehicle or hybrid, (5) use alternative fuels and fuel cells, (6) telecommute from home, and (7) reduce air travel.[811] America will increase its current population of 300 million to a population of 400,000 by the year 2040.[812] The number of cars on American roads is expected to double by 2020.[813] The energy to power more than 400 million personal automobiles and another 100 million motor vehicles to deliver goods and services is staggering as is the resulting pollution—including carbon emissions—and unavailable and unaffordable road and parking capacity. On average, each person in the U.S. travels 40 miles daily on land and 5 miles daily by airplane amounting to 60% more daily mileage than 30 years ago.[814] Commuting only accounts for 20% of all trips; the growth in travel demand is taking place in non-work based activities such as social, shopping, and recreational purposes.[815] Although the U.S. is home to only 5% of the world's population, its residents own almost a third of the world's cars which account for 45% of the carbon dioxide emissions generated by cars

---

811. Gore, *supra* note 4 at 311–13.
812. Lang & Nelson, *supra*, note 711.
813. Tamminen, *supra* note 148 at 71.
814. Hillman et al, *supra* note 101 at 49.
815. Banister, *supra* note 170. Unsustainable Transport: City Transport in the New Century 124 (2005).

Traffic, Los Angeles. Source: Richard Riesemberg, www.rickrise.com.

worldwide.[816] The annual cost of operating the cars in the U.S. amounts to $1.444 trillion dollars.[817] Also, the opportunity costs of the work force in the automobile manufacturing, sales, servicing, parts, supplies, and repair industries rather than working in sustainable transport and living must be added to the operating costs.

In 2002, the U.S. spent $200,000 per minute, about $105 billion per year, on foreign oil.[818] The United States currently spends $612,500 per minute for the 60% of the 21 million barrels of foreign oil consumed daily.[819] The average American consumes 69 barrels of oil annually.[820] To make matters worse, since 1987, the average miles-per-gallon of vehicles has dropped due to the market demand for SUVs, light trucks, and heavy sedans. People are living

---

816. Ewing et al., *supra note* 632 at §1.3.
817. Richard Register, Ecocities: Rebuilding Cities in Balance With Nature 143 (Rev. ed. 2006).
818. Black, *supra* note 484 at 268.
819. Tamminen, *supra* note 148 at 73.
820. Hiro, *supra* note 462 at 331.

Hydrogen-fueled bus, Estoril, Portugal. Source: Carfree.com.

further out in a pattern of sprawl so that miles traveled per person doubles every 10 to 15 years.[821] Automobiles are also toxic for their inhabitants, containing carcinogens and poisonous phthalates and other particles that can exacerbate allergies and asthma and even reduce penis size and retard genitalia development in children.[822]

Sustainablity and survivability demands a different system. In addition to converting energy to renewable forms, transport is a most significant component of carbon emissions. The 400 million motor vehicles in the United States are not the only source of transportation carbon emissions; air transport, sea transport, and trucking, together with construction and farm equipment generate significant carbon emissions. At the end of the last century 26% of the U.S. energy consumption was from transportation.[823] Although industrial and residential consumption are declining, transportation-based energy consumption is now at

---

821. TAMMINEN, *supra* note 148 at 71.
822. Michael Abrams, *Invisible Hitchhikers: Slide Behind the Wheel and your Biggest Health Threat May Not be the Other Drivers On the Road, but the Invisible Toxins Riding Shotgun*, MS.COM, Mar. 15, 2008, *available at* http://health.msn.com/health-topics/asthma/articlepage.aspx?cp-documentid=100185275&GT1=31005.
823. Berger, *in* CLIMATE CHANGE POLICY 411, 434, *supra* note 428.

37%[824] and transportation accounts for 40% of carbon emissions in the United States.[825] The transportation share of energy use is anticipated to increase to 42% by 2025, accounting for half of the 40% increase in U.S. carbon emissions in 2025.[826] The 225 million automobiles in the U.S. is increasing by 3 million annually averaging 1.2 private vehicles for each driver whose annual car mileage has grown by 50% since 1970.[827] The internal combustion engine consumes 63% of the nation's petroleum use.[828] The automobile burns fuel derived from 100 times its weight in ancient plants daily; yet, a mere 0.3% of that fuel moves the driver.[829] A British study found that truck transport generates 180 grams of carbon per ton per kilometer; trains generate 15 grams, a saving of 92%; but only 12% of freight is transported by rail.[830] A round trip air flight from Washington, D.C. to New York, a trip of 480 miles, generates 2.21 metric tons of carbon dioxide while the same trip by car generates 0.18 metric tons.[831] The trip by train generates only 0.15 metric tons which, when divided by the number of passengers, shows that rail travel or bus travel generates the least emissions.[832] Measuring the energy intensity of travel modes, excluding walking and cycling, indicates that demand-response taxis and vans are the least efficient expending 14,301 BTU per passenger mile with the most efficient mode being vanpooling at 1,294 BTUs.[833] Carpooling with four people generates 1,840 BTUs, while cars with only a driver were rated at 7,380.[834] Rail expends 710 BTUs, and a trolley around 1,050 BTUs.[835] Road and rail freight have increased by 56% and 30% respectively during the past 10 years while transportation by water, the least carbon intensive, dropped by 12%.[836]

---

824. HILLMAN ET AL, *supra* note 101 at 44.
825. KUNSTLER, *supra* note 99 at 118. *See also* EWING ET AL., *supra* note 632 at §§ 1.1, 2.1 (transportation sector accounts for 28 percent of total greenhouse gas emissions in the U.S. and 33 percent of the nation's energy-related carbon dioxide emissions).
826. HILLMAN ET AL, *supra* note 101 at 66, 69.
827. *Id* at 50–51.
828. BLACK, *supra* note 484 at 261.
829. IAIN CARSON & VIJAY V. VAITHEESWARAN, ZOOM: THE GLOBAL RACE TO FUEL THE CAR OF THE FUTURE 305 (2007), *citing* Amory B. Lovins, *Winning the Oil Endgame: Innovation for Profits, Jobs, and Security*, 84 FOREIGN AFFAIRS 152 (2004).
830. MONBIOT, *supra* note 11 at 147.
831. HILLMAN ET AL, *supra* note 101 at 218–20.
832. *Id.*
833. *Driving to Green Buildings, supra* note 639.
834. Marcia D. Lowe, *Cars, Their Problems, and the Future*, in CITIES AND CARS 221, 222 (Roger L. Kemp ed. 2007).
835. *Id.*
836. HILLMAN ET AL, *supra* note 101 at 58.

The rate of increase in automobiles in China is staggering. From no automobiles on the road in 1990, to 10 million automobile owners in 1995, and to 27 million in 2004[837] a growth pattern anticipated to continue adding another 6 million vehicles each year.[838] By 2020, China's annual consumption of gasoline will equal that of the United States.[839] China's increase in oil consumption is 16% annually.[840] Although new cars in India numbered 200,000 in 1995, sales reached one million in 2005, a 500% increase in a decade. India currently has 8 million automobiles on the road and 40 million citizens able to purchase a car.[841] The auto industry's 60 million vehicles produced each year consume the lion's share of the 85 million barrels of oil produced daily.[842] By the end of this decade, 900 million vehicles will be operating worldwide and by 2050 the number is projected to rise to 1.5 billion vehicles.[843] Globally, car ownership is anticipated to increase by 75% and traffic is anticipated to grow by 56% by 2020.[844] Even if miraculously these vehicles were all powered by hydrogen, the production of hydrogen cells and infrastructure would require increasing electrical output by 15 to 20%.[845] The only technology that could achieve this surge without generating enormous additional carbon emissions would be nuclear power.[846] Although the precise date for reaching, or having reached, peak oil[847] (when the supply of oil begins to decline) is unknown, it is clear that the future is linked to alternative fuels, propulsion, or power sources, and most realistically, alternative modes of transport.[848] Peak oil[849] has been reached or soon will be with the dramatically increasing number of motor vehicles through-

---

837. Hiro, *supra* note 462 at 194–95.
838. Tamminen, *supra* note 148 at 74–76.
839. *Id* at 74–76 (China will add nearly sixty million more vehicles by 2010 and as it reaches American consumption levels will require the supply consumed by the entire world today).
840. Hiro, *supra* note 462 at 195.
841. *Id* at 231–32.
842. Carson & Vaitheeswaran, *supra* note 829 at 21.
843. Hiro, *supra* note 462 at 288–89.
844. Banister, *supra* note 170 at 20.
845. Hiro, *supra* note 462 at 288–89.
846. *Id.*
847. Kunstler, *supra* note 99 at 24–8; Pahl, *supra* note 99 at ix–xxv; Tertzakian, *supra* note 99.
848. *See* Brown, *supra* note 36 at x, 21–40 (projecting that China will require 99 million barrels of oil daily by 2031, compared to the current world production of 84 million barrels); Goodstein, *supra* note 498 at 17–54.
849. Kunstler, *supra* note 99 at 24–8; Pahl, *supra* note 99 at ix–xxv; Tertzakian, *supra* note 99.

Tram in Bratislava, Slovakia.
Source: James A. Kushner.

out the world.[850] One of the problems with the imbalance between automobiles and public transport and congestion is that drivers are provided free roads and free parking and landowners, including the carless, must pay for it.[851]

A serious current problem that inhibits rational transportation policy is that sprawl and automobile transport is heavily subsidized.[852] The fossil fuel industry receives more than $210 billion in subsidy annually, having supported political campaigns with $181 million between 1990 and 2004.[853] In addition, a study by Mark Delucchi at the U.C. Davis Institute for Transportation Studies found that U.S. drivers are underpaying local, state, and federal governments by $40 billion to $105 billion annually and that fuel tax and fees fail to

---

850. PITTOCK, *supra* note 36 at 167–68. *See also* RICHARD HEINBERG, POWER DOWN: OPTIONS AND ACTIONS FOR A POST-CARBON WORLD 17–54 (2004).

851. SHOUP, *supra* note 171; Michael Lewyn & Shane Cralle, *Planners Gone Wild: The Overregulation of Parking*, 33 WM. MITCHELL L. REV. 613 (2007); Rachel Weinberger, *The High Cost of Free Highways*, 43 IDAHO L. REV. 475 (2007).

852. *See generally* Michael Lewyn, *How Government Regulation Forces Americans into their Cars: A Case Study*, 16 WIDENER L.J. 839 (2007).

853. BROWN, *supra* note 36 at 77–78.

pay the costs of roads and highways by 20 to 70 cents per gallon of gasoline, excluding environmental and uncompensated crash costs.[854] General revenues and property taxes were required to cover road costs in the amount of $700 million annually in New Jersey, and $2 billion annually in New York.[855] Between $30 billion and $60 billion is spent annually for military expenditures protecting access to Middle Eastern oil while $257 billion, or roughly $2,000 per taxpayer annually, subsidizes automobile use.[856] The Iraq war could cost between $1 trillion and $2 trillion.[857] Another way to view the gas subsidy is that the actual costs, including social costs, to the United States is $11 per gallon.[858] These actual costs should be used to establish a realistic gas tax.[859] The federal gas tax has been at 18.4 cents a gallon since 1993 which is approximately 3.7 cents per mile if a car gets 13 miles per gallon and less than a penny per mile for a hybrid that gets 44 miles per gallon.[860] Improved gas mileage will deplete the federal Highway Trust Fund that funds highway infrastructure and transit funding will decline from an $8.1 billion surplus in 2007 to a $1.7 billion deficit by 2009.[861] This would explain why the government is encouraging toll roads, toll lanes, and congestion pricing as an alternative to a realistic carbon tax charged by the mile rather than the gallon.[862] Mary Peters, the Secretary of Transportation in the Bush Administration opposes the proposal of the National Surface Transportation Policy and Revenue Commission to increase the gas tax to 40 cents per gallon or any increase in the gas tax which she consid-

---

854. *Delucchi Study Finds that U.S. Motorists Do Not Pay Their Way*, STREETSBLOG, Sept. 20th, 2007.
855. *Id.*
856. BROWN, *supra* note 36 at 77–78.
857. MONBIOT, *supra* note 11 at 56; TAMMINEN, *supra* note 148 at 59 ($1 trillion, $100 billion annually). *See also* Mark Mazzetti & Joel Havemann, *Iraq War is Costing $100,000 per Minute*, L.A. TIMES, Feb. 3, 2006 (Afganistan War costs $18,000 per minute); Martin Wolk, *Cost of War Could Surpass $1 Trillion*, MSNBC, Mar. 17, 2006, available at http://www.msnbc.msn.com/id/11880954/ (cost growing at $200 million per day).
858. BROWN, *supra* note 36 at 231.
859. *Id.*
860. Bruce Siceloff, *Drivers Might Pay Road Taxes by Mile*, NEWS OBSERVER, Jun. 17, 2007, available at http:www.newsobserver.com/news/growth/v-print/story/607113.html.
861. *Id.*
862. Michael H. Schuitema, Comment, *Road Pricing as a Solution to the Harms of Traffic Congestion*, 34 TRANSP. L.J. 81 (2007). *See also* Lyndsey Layton & Spencer S. Hsu, *Letting the Market Drive Transportation: Bush Officials Criticized for Privatization*, WASHINGTONPOST.COM, Mar. 17, 2008, *available at* http://www.washingtonpost.com/wp-dyn/content/article/2008/03/16/AR2008031603085.html.

ers already too high.[863] Even with gas subsidies, drivers pay $34 per 100 miles to drive compared to $14 if they traveled by public transport.[864] The total federal annual subsidy for trains equals less than the cost of replacing 1.5 miles of Interstate 880 damaged by the 1989 Loma Prieta earthquake.[865]

> A cost of somewhere between $127 billion and $374 billion a year for offstreet parking has been shifted into higher prices for everything else. This cost disappears from sight when drivers park free, but it does not cease to exist. Instead, free parking increases the demand for driving, which in turn increases the subsidy necessary to meet the peak parking demand. Minimum parking requirements are truly a great planning disaster—perhaps the greatest of all time.[866]

In addition to fuel subsidies, most communities mandate that developers of residential, commercial, office, and industrial projects provide free parking. The area of land in the U.S. devoted to parking is 4,950 square miles, about the size of Connecticut and larger than Delaware and Rhode Island combined.[867] The annualized capital and operating cost of off-street parking in the U.S. is between $79 billion and $226 billion.[868] By comparison, the annual cost of operating roads and highways is between $98 billion and $177 billion. The average monthly cost of a parking space is $127, representing a $5.77 subsidy to the commuting worker.[869] While drivers pay $3 billion annually for parking, the rest of the cost is bundled in the price of goods, services, and housing and thus 96 to 99% of the annual cost of parking is a subsidy.[870] Drivers pay at most 4% of the cost of parking. When drivers paid 8.4¢ per mile in 1990 for automobile operation, the subsidy for free parking was as high as 11¢ a mile, as much as 131% of the costs of operation.[871] The free parking subsidy, particularly when considering free parking at stores and at home where government mandated that the developer provide parking infrastructure as a condition of development, represents a subsidy that would be greater than if the government

---

863. Mary E. Peters, *Gas Taxes Are High Enough*, WALL ST. J., Jan. 18, 2008, at A13, *available at* http://online.wsj.com/public/article_print/SB120062474267899727.html.
864. Lowe, in CITIES AND CARS, *supra* note 834 at 223.
865. REGISTER, *supra* note 817 at 153–54.
866. SHOUP, *supra* note 171 at 218.
867. *Id* at 217.
868. *Id* at 206.
869. *Id* at 2.
870. *Id* at 206.
871. Id at 207.

provided free gasoline.[872] The total social cost of free off-street parking is between $127 and $374 billion annually, as much as the peacetime national defense or Medicare budgets.

To offset the parking subsidy, gasoline taxes would have to be raised as much as $3.74 per gallon.[873] The value of parking in the U.S. is more than the value of all vehicles and twice the value of all roads.[874] Free parking reduces the cost of the commute by 71% or $5.77 daily.[875] Removing the parking subsidy and adding it to the price of a gallon of gas through a parking tax would increase the price by $4.44.[876] In addition, the cost of providing free parking has discouraged developers from constructing multi-family housing or redeveloping older buildings, due to the rent-inflating cost of parking regulations and is having an inflationary effect on the rents of the poor and working class. Minimum parking requirements add as much as 38% to the cost of apartment development.[877] The poor, the sector of society that lives most sustainably and often relies on public transport, must subsidize the drivers as the poor ultimately pay for free parking through higher prices for goods, services and rents.[878] In cities other than Los Angeles, which typically offers occupants free parking, the amount of parking is being restricted and drivers must purchase a parking space that can cost as much as $225,000 in Manhattan.[879] A New York City parking space exceeds not only the cost of the average American home, in cost per square foot it often exceeds the price of the owner's finished condominium.[880] Allowing the market to set the price for parking and limiting parking availability will educate the public on the true cost of the automobile and swiftly shift travel preference to public transport, biking and walking.

Sadly, while the Bush administration began to fund demonstration projects for toll roads and congestion pricing (imposing charges for entering con-

---

872. Lewyn & Cralle, *supra* note 851 at 617.
873. SHOUP, *supra* note 171 at 207–08.
874. *Id* at 209–211.
875. *Id* at 212.
876. *Id* at 214.
877. *Id* at 148–51.
878. *Id* at 165 n. 1.
879. *Manhattan Parking Spot Going for $225,000*, CNNMONEY.COM, July 12, 2007 (indicating prices in Boston can sell for as much as $175,000 and in Chicago as much as $75,000).
880. *Id.* (indicating that the average parking space in Manhattan costs $1,100 per square foot, while the average apartment rents at $1,107 per square foot; while a two bedroom unit in the building charging $225,000 for a parking space sells for $2.2 million, that equates to $1,281 per square foot while the parking space costs $1,500 per square foot).

Double articulated bus on dedicated busway at Utrecht University in The Netherlands. Source: James A. Kushner.

gested cities to discourage driving), it also began a program to subsidize worker parking.[881] Mary Peters, the Secretary of Transportation in the Bush Administration opposed spending federal money on bike paths as she did not consider bicycles to be transportation.[882] A visit to the Netherlands where bicycles outnumber automobiles making the nation quite pleasant may have convinced her otherwise. As of 1990, federal transportation funds for walking and cycling amounted to $2 million annually.[883] Walking constitutes 0.7% and cycling 0.2% of person-miles of daily travel in the U.S.[884] Half of all students go to school by car, but if just 6% of those driven walked or cycled, saving 1.5 million dropoffs and pickups, 60,000 gallons of gasoline would be saved daily.[885]

---

881. William Neuman, *Mixed Signals: Driving to Work as a Tax Break*, N.Y. Times, Aug. 16, 2007.

882. *Secretary Peters Says Bikes "Are Not Transportation,"* Excerpt from Interview on PBS "NewsHour" with Jim Lehrer, *available at* http:www.streetsblog.org/2007/08/17/secretary-Peters-says-bikes-are-not-transportation/.

883. Hillman et al, *supra* note 101 at 37.

884. *Id* at 55.

885. Rogers & Kostigen, *supra* note 209 at 42.

Bicycles, Amsterdam, NL. Source: James A. Kushner.

The 60,000 students, or 2.5% of U.S. students that cycle to school save 100,000 gallons of gas daily.[886] Under the initiative established in 1997 legislation, U.S. employees can spend pre-tax wages up to $215 a month to pay for parking at work or at a commuter rail station, amounting to approximately a $1,000 annual subsidy designed to make driving to work more attractive.[887] The provision also allows pre-tax dollars to be spent on transit but that provision is usable largely by suburban commuters using expensive rail travel.[888]

Worldwide, the annual subsidies for burning fossil fuel is $700 billion[889] and $1.7 trillion is spent on purchasing oil annually.[890] Air travel is the least sustainable transport and the fuel it uses is statutorily exempted from taxation thus creating an incentive to pollute.[891] While planes are becoming more efficient, the huge growth in passenger miles results in escalating carbon emis-

---

886. *Id* at 43.
887. Neuman, *supra* note 881.
888. *Id*.
889. Brown, *supra* note 36 at 233.
890. *Id* at 248.
891. 3 Johansen, *supra* note 1 at 677–79.

sions.[892] In 1992, air travel generated 2% of carbon emissions and 10% of fuel use in the U.S.[893] U.S. domestic air travel is expanding at 2% annually and is expected to double by 2020.[894] Each passenger mile emits 0.64 pound of carbon dioxide.[895] Were jets to fly lower, cloud formation carbon emissions would be cut by 4%.[896] Due to water vapor in jet contrails, jets actually mask global warming for when planes were grounded between September 11 and 14, 2001 daytime temperature abruptly increased by 2º F[897] suggesting that jets should continue flying until carbon emissions are drastically reduced by adopting alternative ground and sea transport propulsion.[898] Until air travel is eliminated, limiting passenger travel to necessary trips and strictly limiting baggage weight will help reduce emissions. Other than air travel for a limited air defense, air travel cannot continue at the present rate given the carbon emissions, the total lack of an alternative power source technology, and the future unavailability of gasoline.[899] The supersonic jet aircraft is the most environmentally damaging air transport technology ever developed and there will never be an eco-friendly F-35 Joint Strike Fighter.[900] The take-off of a Boeing 747 is the equivalent of setting a gas station on fire and flying it over the neighborhood or running 2.4 million lawn mowers for 20 minutes.[901] Taxing air travel, eliminating budget flights, and charging passengers by weight will also apportion actual costs and restrict excessive emissions. Every addition of 10 pounds per traveler in luggage or girth requires an additional 350 million gallons of jet fuel annually, enough to keep a 747 flying continuously for 10 years.[902] Unfortunately, sea travel in fossil fuel-powered ships for commerce and recreation is also unsustainable. Some cruise ship lines, however, may be showing the way toward wind-powered transport. Currently, 80% of the world's international trade in goods are transported by ship.[903] The future of intercontinental travel may be sailing, particularly integrating the technical knowledge from the wind power

---

892. *Id.*
893. FLANNERY, *supra* note 4 at 282.
894. HILLMAN ET AL, *supra* note 101 at 68.
895. ROGERS & KOSTIGEN, *supra* note 209 at 143.
896. FLANNERY, *supra* note 4 at 283.
897. *Id.*
898. *Id.*
899. MONBIOT, *supra* note 11 at xiii.
900. *Id* at 60.
901. Jerry Mander, *Globalization is Harmful to the Environment*, *in* GLOBALIZATION: OPPOSING VIEWPOINTS 84, 88 (Louise I. Gerdes ed. 2006).
902. ROGERS & KOSTIGEN, *supra* note 209 at 25.
903. Mander, *in* GLOBALIZATION, *supra* note 901 at 87.

Smart Car, Wurtzburg, Germany. Source: James A. Kushner.

industry and technology development with the engineering of shipping. One hope for replacing air, rail, truck, and sea transport is travel and transport by helium blimps powered by hybrid engines.[904]

The air pollution and smog as well as carbon emissions from motor vehicles that exacerbate global climate change are also a public health hazard.[905] Nationally, automobile pollution costs $24.3 billion annually.[906] In the Los Angeles area, transportation sources account for between 60 and 90% of all air pollution.[907] The annual cost in increased health care caused by the effects of petroleum just in California amounts to between $9 billion and $240 billion as compared to $8 billion in tobacco-related costs.[908] Nationally, direct annual health care costs are between $54.7 billion and $672.3 billion.[909] California's direct state health care payments related to petroleum pollution are between $500 million and $12 billion.[910] Petroleum-based diseases in California result

---

904. *Technology: Don't Call it a Blimp*, 210 NAT'L GEOGRAPHIC, Nov. 2006, at 14.
905. TAMMINEN, *supra* note 148 at 27–51.
906. *Id* at 57.
907. BLACK, *supra* note 484 at 262.
908. TAMMINEN, *supra* note 148 at 54.
909. *Id* at 57.
910. *Id* at 54.

in 9,300 deaths, 16,000 hospital visits, 600,000 asthma attacks, and 5 million lost work days annually.[911] In addition and in California alone, petroleum causes 1.7 million cases of respiratory illness and 1.3 million school absences.[912] It is time to declare both petroleum, internal combustion engines, and carbon a public nuisance, ban the use of fossil fuels, and bring private and public lawsuits for damages as in the case of tobacco.[913] Petroleum also generates $150 million in crop losses a year just in California's San Joaquin Valley, and statewide a loss of $300 million in farm income reflecting a loss of $54 million in federal income taxes that would have been paid.[914] Nationally, crop losses from motor vehicle pollution is between $3 billion and $6 billion annually.[915] Damage to buildings from motor vehicle pollution across the country is $1 billion to $8 billion annually.[916] Annual forest damage from motor vehicle pollution is $2 million to $2 billion.[917] Water pollution from leaking oil tanks, oil spills, and polluted runoff is between $4 million to $1.5 billion annually.[918] In 1996, road congestion cost the United States $74 billion reflecting 4.6 billion lost hours and 6.7 billion wasted gallons of fuel.[919] The Texas Transportation Institute in its 2007 Mobility Report conservatively estimates that Americans wasted 4.2 billion hours and 2.9 billion gallons of fuel while sitting in traffic in 2006.[920] Anecdotally, congestion is much worse and far more costly today. A serious problem with congestion is the volume of traffic cruising in search of a cheap, on-street parking meter or free parking. Although raising parking meter fees will increase available spaces and reduce cruising, as long as parking is more than $1.20 per hour (the cost of operating a car), it will be cheaper to cruise than to park.

Automobile dependency prevents the redesign of cities and the development of new communities around smart growth, public transport, and New Urbanism. Although scientists are searching for alternative fuels, a true alter-

---

911. *Id* at 55. *See also* BLACK, *supra* note 484 at 262 (citing 2.8 million lost workdays according to the state's air resources board).
912. BLACK, *supra* note 484 at 262.
913. TAMMINEN, *supra* note 148 at 191–202.
914. *Id* at 55–56.
915. *Id* at 57.
916. *Id.*
917. *Id.*
918. *Id.*
919. BLACK, *supra* note 484 at 263.
920. Patrik Jonsson, *Planners Raise Local Funds for Innovative Projects Instead of Relying on State and Federal Money*, CHRISTIAN SCIENCE MONITOR, Nov. 27, 2007, available at http://www.csmonitor.com/2007/1127/p03s03-usgn.htm.

Bike Rental, Vienna, Austria. Source: James A. Kushner.

native does not currently exist. Ethanol, designed to reduce gasoline consumption has caused competition with food and higher food prices when produced from corn.[921] Under current practice, ethanol requires the use of gas in farming and large amounts of water. One gallon of Ethanol requires 1.29 gallons of petroleum to produce.[922] The land necessary to produce 40 bushels of corn yielding 100 gallons of ethanol worth $200 could produce $100,000 worth of wind energy each year.[923] Ethanol from corn generates twice the carbon emissions of gasoline when the land use changes needed for increased corn production are considered.[924] The use of sugar cane rather than corn for ethanol would yield 650 gallons rather than 350 gallons per acre; the net energy yield

---

921. Michael S. Rosenwald, *The Rising Tide of Corn*, WASH. POST, Jun. 15, 2007, at D01. *See also* BLACK, *supra* note 484 at 286–92 (ethanol a scam). *But cf.* Msangi & Rosegrant, *supra* note 117 (suggesting that biofuel development carries economic opportunities that may outweigh food versus fuel debates with perhaps excessive optimism for environmental impacts and problems of water and emissions).

922. BLACK, *supra* note 484 at 286–87.

923. BROWN, *supra* note 36 at 200.

924. Associated Press, *Study: Corn Ethanol No Climate Solution: Greenhouse Emissions Much Higher if Land Use Factored In, Researchers Say*, MSNBC.COM, Feb. 7, 2008, available at http://www.msnbc.msn.com/id/23057867/.

is 8 for sugar cane compared to only 1.5 for corn.[925] Ethanol is receiving a deep subsidy in the form of a $2.1 billion annual tax credit worth 51 cents per gallon under the American Jobs Creation Act of 2004.[926] The excessive subsidy for both corn and corn-based ethanol have increased corn prices by up to 40%, generated food riots in Mexico as tortilla prices have increased by 60%, and have stimulated deforestation around the globe to plant the more valuable corn crop.[927] Estimates predict that the ethanol policy of the United States will cost $1.38 per gallon of subsidies, will replace but 1.5 million barrels of gasoline each day or 7% of daily needs, and will cause 600 million people worldwide going hungry by 2025 due to the rising food costs generated by corn-ethanol subsidies.[928] In addition, strikes and riots are occurring around the world as food for energy is driving up prices with pasta in Italy up by 20% as wheat has doubled in price in a decade.[929] Most of the 100 million tons of cereal grain which is nearly all corn amounting to 12% of the corn consumed on the planet has gone to biofuel resulting in rising prices of food and a 5% decline in food stocks.[930]

Palm oil, another alternative, would yield 500 gallons of biodiesel per acre compared to 56 gallons per acre from soybeans.[931] Unfortunately, palm oil and sugar cane are grown in tropical areas and are likely to generate deforesta-

---

925. BROWN, *supra* note 36 at 200; Jeff Goodell, *The Ethanol Scam: One of America's Biggest Political Boondoggles*, ROLLING STONE.COM, July 24, 2007, available at http://www.rollingstone.com/politics/story/15635751/the_ethanol_scam_one_of_americas_biggest_political_boondoggles (8:1 energy to fossil fuel for sugar-based ethanol, gasoline at 5:1, and corn-based ethanol at but 1.3:1). *See generally* Jocelyn D'Ambrosio, Student Article, *Alternative Fuels: An Evaluation of Corn Ethanol, Cellulosic Ethanol, and Gasoline*, 37 ENVTL. L. REP. NEWS & ANALYSIS 10615 (2007) (although corn ethanol will reduce emissions somewhat, Cellulosic Ethanol would be significantly more efficient, produce more fuel, and reduce emissions).
926. PUB. L. No. 108-357, 118 Stat. 1418 (2004), 26 U.S.C.A. 896–907 (2007). *See* BLACK, *supra* note 484 at 286–87.
927. Goodell, *The Ethanol Scam, supra* note 925.
928. *Id.*
929. J. Michael McConnell, *Annual Threat Assessment of the Director of National Intelligence*, Feb. 5, 2008, at 44, *available at* http://www.tsa.gov/assets/pdf/02052008_dni_testimony.pdf (statement of the Director of National Intelligence to the Senate Select Committee on Intelligence reporting corn protests in Mexico, bread riots in Morocco, and unrest in Burma); Peter Gumbel, *Pasta Panic: The Price of Wheat is Up 60% This Year, and in Italy They're Taking to the Streets Over the Cost of Tortellini*, 156 FORTUNE 47 (2007).
930. Lederer, *Biofuel Growth Adds to Hunger, supra* note 119.
931. BROWN, *supra* note 36 at 201.

Richshaw bicycle, Copenhagen, Denmark. Source: James A. Kushner.

tion.[932] Another scheme to extend the oil supply is to convert oil sands and shale to petroleum. Canada's Albian Sands a 140,000 square kilometer mine in Alberta, is the largest identified sand, shale, or liquid oil reserve outside of Saudi Arabia, equal to the size of the state of New York.[933] Strip mining of the coral sands, however, would destroy wildlife and habitat of the world's largest remaining unspoiled forest, contributing significantly to global warming and generating five times the carbon emissions involved in extracting oil.[934] Extracting Alberta, Canada's oilsands, a field the size of Florida, has been called the "most destructive project on Earth.[935]

Hydrogen cells are often mentioned as an alternative to oil, but hydrogen is dangerous and difficult to handle and store and hydrogen cells, like batteries, are not a source of energy but only a storage and transporting mechanism.[936] Hydrogen is particularly dangerous and highly flammable as it burns

---

932. Lester R. Brown, Plan B 2.0: Rescuing a Planet Under Stress and a Civilization in Trouble 201 (2006).
933. Sheila McNulty, *Green Leaves, Black Gold*, Financial Times, Dec. 16, 2007.
934. Id.
935. *Feds Allowing Tarsands to Become 'Most Destructive Project on Earth': Report*, CBC News, Feb. 15, 2008.
936. Goodstein, *supra* note 498 at 38–39.

with an invisible flame, is leak prone, and can be ignited by an electrical storm or a spark from a cell phone.[937] Hydrogen technology actually generates more carbon emissions than oil and its cost is prohibitive. To fill all the cars in the U.S. with hydrogen would require four times the current capacity of the national grid.[938] A 40-ton truck could only deliver 100 gallons of compressed hydrogen which would require 15 trucks to deliver what is currently contained in a 26-ton gasoline tanker.[939] Such a transport system would consume 40% of the fuel carried.[940] The hydrogen tank on a vehicle would have to be ten times the size of a gas tank and a significant percent of the fuel would boil-off or evaporate daily.[941]

Hydrogen is but a trace element in our atmosphere comprising half a part per million. With more production, storage, and use, leaks would likely result, increasing the atmospheric concentration and secondarily increase the abundance of methane by up to 4%, exacerbating greenhouse gas concentration.[942] Hydrogen leakage would be likely if cells were widely used posing the serious threat of worsening stratospheric ozone depletion.[943] If renewable energy sources were used to develop hydrogen cells, it would be wasteful since the same energy could be used to generate power.[944] In addition, hydrogen is dangerously combustible and highly corrosive; a new pipeline, storage, and transport technology would have to be developed at high cost and the use of gasoline to transport hydrogen would further make it uneconomical.[945] Experimental hydrogen vehicles can be extremely expensive to operate with Zero-emission buses costing more than $50 per mile as compared to diesel at less than $2 per mile.[946] Hydrogen is receiving a lot of research money, while more promising

---

937. FLANNERY, *supra* note 4 at 264.
938. MORRIS, ENERGY SWITCH, *supra* note 498 at 143.
939. FLANNERY, *supra* note 4 at 263.
940. *Id.*
941. *Id.*
942. *Id* at 293.
943. 3 JOHANSEN, *supra* note 1 at 690–91.
944. ROMM, HELL AND HIGH WATER, *supra* note 10 at 186.
945. KUNSTLER, *supra* note 99 at 110–16.
946. Gary Richards, *VTA Finds Hydrogen Buses Cost Much More to Run Than Diesel Vehicles*, MercuryNews.com, Feb. 26, 2008, *available at* http://www.mercurynews.com/ci_8365544 (San Jose newspaper reporting per mile cost of operation for the Valley Transit Authority at $51.66 as compared to $1.61 for conventional diesel buses, a cost of $2.5 million as compared to the $400,000 price of a diesel bus, and requiring repairs after 1,100 miles as compared to 6,000 for a diesel bus while reporting that AC Transit in the East Bay has had more efficient results from hydrogen-hybrid buses that also have electrical batteries).

technologies are denied research funding.[947] General Motors has spent more than $1 billion on hydrogen research but has generated no prospect for any profit in the foreseeable future.[948] Hydrogen development faces a chicken and egg problem in that hydrogen-fueled cars will not be produced without filling stations and the oil companies, as in the case of ethanol and bio-fuels, will not establish stations until those automobiles have been produced. The oil companies are not likely to act until gasoline is unavailable or excessively costly.[949] Were hydrogen produced from natural gas at the gas station, 50% more carbon emissions would result compared to using the natural gas to power vehicles.[950] Hydrogen vehicles are unlikely to achieve even a 5% market penetration by 2030 and will have no meaningful commercial success in the near- or medium-term.[951] The cost of hydrogen energy will be at least four times the cost of gasoline and a technological breakthrough will be required to solve the storage problem.[952] A promising project, however, is Honda's FCX hydrogen cell car that would be filled at home in a home-based energy station that will fuel the vehicle cell and provide energy to power the home and is to be available by 2010.[953]

Hybrid cars with an additional storage battery and a plug-in capacity could reduce gasoline use by 70%.[954] If built using advanced polymer composites, they could reduce gasoline use by 85%.[955] Existing technology, such as the vehicle which uses hybrid lithium-ion batteries with plug in recharging ability rather than cheap, heavy, and short-lasting lead-acid batteries at an additional cost of $12,000 and uses the internal combustion engine only at high speeds, results in a savings of up to 100 miles per gallon.[956] Converting to hydrogen or hybrid cars would do little for traffic congestion, however, and the urban de-

---

947. ROMM, HELL AND HIGH WATER, *supra* note 10 at 185–88. *See also* ROMM, THE HYPE ABOUT HYDROGEN, *supra* note 60 at 3–4 (not feasible and available technology utilizes fossil fuel for production).
948. CARSON & VAITHEESWARAN, *supra* note 829 at 15.
949. BLACK, *supra* note 484 at 304–05.
950. FLANNERY, *supra* note 4 at 263.
951. ROMM, THE HYPE ABOUT HYDROGEN, *supra* note 60 at 9. *See also* BANISTER, *supra* note 170 at 158–68; HEINBERG, *supra* note 850 at 125 ("the idea of a full-blown "hydrogen economy" is probably more hype than reality); HIRO, *supra* note 462 at 282–89 (hopefully in 2050).
952. BANISTER, *supra* note 170 at 168.
953. BLACK, *supra* note 484 at 293–316.
954. *Id.* at 192.
955. *Id* at 192–93.
956. CARSON & VAITHEESWARAN, *supra* note 829 at 15, 268–272.

sign effects of the automobile such as sprawl, would continue to require the personal automobile for transport. Currently, hybrids are projected to increase gas milage by 25–30%.[957] Hybrids and e-hybrids, i.e., hybrids that can plug into the power grid, are the likely candidate for personal vehicles if the automobile is to be retained.[958] Despite the drawbacks even from relatively clean and relatively less gasoline consumption, shifting to improved hybrids would nevertheless constitute the biggest gain in carbon emissions reduction.[959] Hybrids, which currently sell for from $2,000 to $6,000 more than gasoline-powered cars, reduce pollutants by 90% and carbon dioxide by 50%.[960] To avert the worst effects of global climate change, the planet must reduce carbon emissions by 70% by 2050, a goal that could be attained if every automobile and truck was traded for a hybrid.[961] Some argue that a reduction of 80%[962] or 90%[963] is an essential goal by 2050. By comparison, the Kyoto Protocol set a 5.2% reduction by 2012.[964] Unfortunately, these estimates are based on a stable level of carbon in the atmosphere.[965] Given the anticipated expansion of coal-fired power plants, "even if every SUV were downsized to a Schwinn, every truck and bus repowered to burn biodiesel, and every refrigerator retrofitted to run with solar panels, we are playing Russian roulette with the very thing that makes our life on earth possible—a steady, temperate climate."[966]

Fuel efficiency in transportation can be increased sevenfold with available technology.[967] For example, fuel economy increases 5.6 miles per gallon for each reduction of 1,000 pounds, thus switching from light trucks, SUVs and mid-sized cars to smaller vehicles could easily increase gas mileage by 50%.[968] Shifting to diesel engines can yield up to 60% more fuel economy, but diesels generate more smog.[969] Cutting road speeds can yield significant increases in

---

957. TERTZAKIAN, *supra* note 99 at 193.
958. ROMM, THE HYPE ABOUT HYDROGEN, *supra* note 60 at 201–08.
959. BROWN, *supra* note 36 at 201.
960. HIRO, *supra* note 462 at 280.
961. FLANNERY, *supra* note 4 at 6.
962. MONBIOT, *supra* note 11 at vii (describing the goal of California's Global Warming Solutions Act, Stats. 2006, c. 488 (A.B. 32), *codified at* CAL. HEALTH & SAFETY CODE §§ 38500–38574 (West Supp. 2007)).
963. MONBIOT, *supra* note 11 at vii.
964. GELBSPAN, BOILING POINT, *supra* note 39 at 154; MONBIOT, *supra* note 11 at 48.
965. GOODELL, BIG COAL, *supra* note 8 at 207–08.
966. *Id.*
967. SPETH, *supra* note 431 at 64–65.
968. TERTZAKIAN, *supra* note 99 at 190.
969. *Id* at 191.

Articulated bus, Nuremberg, Germany. Source: James A. Kushner.

gas milage, for a vehicle traveling at 55 miles per hour consumes 17% less fuel than one traveling at 75 miles per hour.[970] Increasing fuel efficiency, however is not the answer to global warming as the history of civilization is that increased efficiency yields increased use and an increase in overall emissions.[971] Healthy cities around the world, those that are consistently ranked as the most liveable and act as a magnet for tourism and economic development, share some common attributes. Each has invested in modern public transport and provides user-friendly, inexpensive, safe and convenient access to destinations. Each has one or more attractive commercial centers which are car-free or traffic-reduced and allow walking from shop to shop with places to eat or a café, tavern, or film to enjoy. These centers can be designed to reflect community values, e.g., as around the proverbial small town model, around ethnic goods and restaurants, or more artistic or fanciful themes such as an Irish, German, Italian, or Korean theme, or one that reflects an historic period. This makes the most sense where a community has a particular type of architectural heritage that can be preserved and enhanced by being woven into to a living community.

---

970. *Id* at 192.
971. MONBIOT, *supra* note 11 at 61–62.

Regardless of the science fiction-like scenario of a magical, technological solution to traffic congestion and global climate change, people and their elected representatives must demand that we pursue a policy of de-autofication. For, regardless of the pollution and energy issues, the increase in automobile and motor vehicle demand as the nation's population moves from 300 million to 400 million by 2040[972] will lead to using land to accommodate traffic that will prevent the development of walkable, pedestrian-friendly, low emissions communities. The space devoted to automobile infrastructure includes between seven and thirteen parking spaces for every vehicle, streets, parking lots, dealerships, repair, paint and body shops, service stations, parts and tire stores, and other car-related activity that could be converted to transit, bicycle, and pedestrian corridors and pedestrian infrastructure. One illustration of wasting space through dependency on automobiles is to consider that an underground metro can carry 70,000 passengers past a single point in a single lane in one hour, while surface rapid rapid rail carries up to 50,000, a trolley or bus in a dedicated lane 30,000, while cars can carry only 8,000 people per hour.[973] These automobile corridors and parking lots can become the high-density, transit-oriented developments to support public transport. With the elimination of automobiles and their infrastructure, more land will be available for parks, landscaped or tree-lined park-like walk and bikeways, urban agriculture, walking trails, dedicated bus routes, or other transit corridors.

1. Local Response

Foremost for local government will be to plan and implement a public transport system that can gradually replace the automobile as the principal means of transport. To make such plans, housing settlements, jobs, commercial and recreational destinations should be planned around transport. In many ways, this will constitute a re-creation of the old streetcar communities. The new transport should include narrower streets and wider sidewalks, walking trails and bicycle paths through parks and quiet streets. In addition to streetcars, there might be trams, buses using dedicated lanes and routes, as well as light rail, subways, or high speed intercity trains.

Transportation is the central issue in sustainable development. Higher density, mixed-use development offers benefits and reduced local trips but may promise only increased traffic congestion and worsened public health unless walk-

---

972. Lang & Nelson, *supra*, note 711.
973. Lowe, in CITIES AND CARS, *supra* note 834 at 222.

ability is teamed with efficient and attractive transit to separate travelers from their cars. Public transport must be a central component of public infrastructure like police and education. This means that it should not be financed from a fare box. It is unfair and environmentally destructive to charge the poor, who comprise the largest component of public transportation patrons and who live the most sustainably, the burden of financing public transit. Transit should be funded from revenues from taxation, particularly taxation of automobile infrastructure such as roads, gas, and automobile purchases and registration. Many transit systems are seeking an easy route to more resources to finance system expansion and higher operating costs through fare increases. Under the Simpson-Curtain rule, for every 10% fare increase, the system experiences a 3.8% drop in ridership.[974] The farebox will destroy transit. The money expended on enforcing fare paying whether barriers or ticket-checking police are extremely expensive. Funds could better be spent on improved service and facilities. Ridership will increase and automobile use and carbon emissions will decline with free fares.

Communities should place a moratorium on development, establish a sustainable plan, and work to establish transit extension. Future tram, light rail, or rail plans in the interim could be established using dedicated rapid bus service which is either fare free or the fare being collected not by drivers, but at kiosks or by cashiers prior to entering the stop so that multiple doors can open and rapid boarding subway-like service can be offered.[975] The buses should stop a minimum of three blocks between stops. Mixed-use, walkable and dense communities should be developed around stops.

The large inventory of detached suburban homes in the United States allows development to be targeted to urbanist, transit-served neighborhoods. New development should be either car-free, traffic free, or traffic reduced.[976] Developers could be offered additional density and tax incentives for developing a car free

---

974. Dave Olson, *Fare-Free Public Transit Could be Headed to a City Near You*, ALTERNET, Aug. 2, 2007, available at http://www.alternet.org/module/printversion/57802. *See also* David Wortman, *No Such Thing as a Free Ride*, SUSTAINABLE INDUSTRIES J., Aug. 3, 2007, available at http://www.printthis.clickability.com/pt/cpt?acrtion=Sustainable+Industries+Journal ($7.50 ferry fare resulted in low ridership in Puget Sound between Kingston and Seattle).

975. MCKIBBEN, *supra* note 339 at ch. 2 (Curitiba, Brazil).

976. KUSHNER, HEALTHY CITIES, *supra* note 738 at 67–73; KUSHNER, THE POST-AUTOMOBILE CITY, *supra* note 170 at 76–84; James A. Kushner, *Car Free Housing Developments: Towards Sustainable Smart Growth and Urban Regeneration Through Car-Free Zoning, Car-Free Redevelopment, Pedestrian Improvement Districts, and New Urbanism*, 23 UCLA J. Envtl. L. & Pol'y 1(2005).

Tram, Freiburg, Germany. Source: James A. Kushner.

development where occupants agree not to own a car. Arrangements with the transit agency might offer a transit pass for occupants of the development. The parking lots and street system can then be converted to gardens, parks, or even wetlands or forest development. Traffic-free communities can be offered less generous incentives but roads, parking, and transit connections such as bus or light rail should be underground so that the surface is safe for children and offers an enhanced amount of green space that can be used for bicycles, pedestrians, parks, and gardens. Higher densities and smaller and affordable units can be encouraged through density bonus program or tax credits or other incentives such as waiving connection, impact or administrative fees. The higher density would support more neighborhood restaurants, bars, coffee houses and shops and would generate a much larger use of public transport.

Traffic-reduced developments should make up the remainder of settlement projects that can be approved. A traffic-reduced development would receive incentives but not as generous as the car-free developments. Transit passes and lower rents and prices would result from the elimination of expensive parking garage spaces. The traffic-reduced development would have narrow roads without parking as well as car-free pedestrian and bike paths and would offer parking at a parking facility on the edge of the development, no closer than the closest transport stop but hopefully at least a sufficient walk to offer some daily exercise. The parking spaces would be sold at the exorbitant price it costs to construct and operate the facility. Revenues can be generated from congestion pricing, such as imposing a fee on driving in transit-oriented development districts, and by raising parking lot taxes and fees for meters.[977]

Commercial areas and residential streets should be narrowed and reduced in speed to below 15 miles per hour to encourage pedestrians and allow outdoor street cafes. In addition to street narrowing, depaving would permit wide gardens buffering housing, shops, or offices from the street.[978] In time, such streets can be closed to traffic and converted to walking and cycling trails.[979] Communities need to plan for a system of equitable reduction in parking and access as additional streets are closed on an annual basis.

Berkeley, California's Measure G seeks to reduce greenhouse gases from the city by 80% by 2050.[980] Under the measure, landlords will provide each tenant with a free transit pass and car sharing vehicles[981] would be parked on every block.[982] All new buildings, resold homes, and renovations would be required to equip garages with outlets for electric cars.[983]

Cities can acquire fleets of vehicles including public transport that run on alternative fuels.[984] London's Heathrow Airport is installing electric driverless pods or Personal Rapid Transport called Ultra Light Transport to carry up to

---

977. Banister, *supra* note 170 at 130–45; Beatley, Green Urbanism, *supra* note 170 at 151–61; Cervero, *supra* note 170; Kushner, The Post-Automobile City, *supra* note 170 at 102–04; Parry & Safirova, *in* New Approaches on Energy and the Environment, *supra* note 170 at 63.

978. Register, *supra* note 817 at 27.

979. *See generally* J. H. Crawford, Carfree Cities (2000); Kushner, The Post-Automobile City, *supra* note 170; Kushner, *Car Free Housing Developments*, *supra* note 976.

980. Jones, *supra* note 315.

981. Beatley, Green Urbanism, *supra* note 170 at 150–56; Kushner, The Post-Automobile City, *supra* note 170 at 99–102; Macht, *supra* note 316; Mannion, *supra* note 316.

982. Jones, *supra* note 315.

983. *Id.*

984. Bulkeley & Bewtsill, *supra* note 313 at 125 (describing Denver).

four persons between terminals taking four minutes as compared to the current 20-minute wait for a bus.[985]

The city of Pasadena, California, along with the state of California, is subsidizing electric bicycles that light rail riders can use to travel between their home and the station so as to avoid the need for park-and-ride lots and the resulting generated traffic.[986] Under consideration is a program to have fleets of clean running vehicles provide door-to door service.[987] Washington, D.C. is expected to adopt legislation that would require commercial landlords to provide bicycle parking for 10% of the available automobile parking spaces.[988]

2. Regional Response

The central role at the regional governance level is to coordinate transit planning and the pace and priorities of line development and extension. Those routes meeting transit oriented development standards should rapidly receive a dedicated bus system. The county or regional governmental unit should also plan to integrate an intercity and highspeed rail system. This will require new track separating freight and passenger rail and laying highspeed tracks between cities within 600 miles of each other. At the regional level, land use controls should be imposed to limit development to a system of polycentric, transit-served neighborhoods, both urban and suburban. A comprehensive transit system should knit the suburbs to each other and to the city.

3. State Response

The state must play a large role in shifting from a car-oriented society to a post-automobile society.[989] For long term solutions, state revenues need to increase and be used to fund public transport and rail systems. The state can mandate local governments to plan for transit and rail improvements, establish a minimum density, maximum parking capacity, and compel mixed-use

---

985. *Arup to Design Infrastructure for Personal Rapid Transport System at Heathrow*, Apr. 11, 2005 (press release), *available at* http:www.arup.com/newsitem.cfm?pageid=7565 (to be operational by 2008).

986. Wortman, *supra* note 974.

987. *Id.*

988. Michael Neibauer, *Legislation to Ease Parking Crunch—for Bicycles*, D.C. Examiner, Sept. 13, 2007.

989. Kushner, The Post-Automobile City, *supra* note 170.

development. States should take the lead in funding trains and local transit service.[990] State tax incentives should also be used to reward those making personal transportation emissions reductions, e.g., by moving to transit-oriented development, or by selling their car. Automobile registration fees should be increased steadily and a personal property tax should be imposed on the value of a motor vehicle, its weight, and the size of its engine. It has been argued that a significant tax on motor vehicles will reduce congestion with an annual benefit to drivers of $1,000 yearly by saving unnecessary fuel, not to mention lost time in traffic,[991]

States can also impose standards on auto makers to reduce emissions or regulate emissions through periodic safety inspections and by requiring minimum miles per gallon efficiency. This is one of the steps California has taken.[992] Maryland is the twelfth state to impose emissions standards on automakers by setting higher mile per gallon limits.[993] The problem with this strategy is there is a good chance the Supreme Court will find state standards to be preempted by lower federal standards or federal control of the subject matter.[994] State par-

---

990. *See* Kamaal R. Zaidi, *High Speed Rail Transit: Developing the Case for Alternative Transportation Schemes in the Context of Innovative and Sustainable Global Transportation Law and Policy*, 26 TEMP. J. SCI. TECH. & ENVTL. L. 301 (2007) (advocating the need for high speed rail in the U.S. as well as in Europe on economic and environmental grounds).

991. Neil Peirce, *Follow British Model on Transportation Needs: It's Time for a Big New Tax in America*, DAILYPRESS.COM, Aug. 19, 2007, *available at* dailypress.com/news/opinion/dp-op_peirce_0819aug19,0,1414318.story (recommending a weight-based tax at the time of sale).

992. A.B.1493, 2001–2002 Sess. (Cal. 2001) (vehicles, including cars and light duty trucks, sold in California beginning with model year 2009 must have 22 percent lower carbon dioxide emissions from the tailpipe than 2002 levels, and then 30 percent lower than 2002 levels by model year 2016). *See generally* Engel, *State and Local Climate Change Initiatives*, *supra* note 172.

993. Lisa Rein, *Md. House Approves Cut in Car Pollution*, WASH. POST, Feb. 21, 2007, at A01 (reporting a 43 mpg state average in 2010 compared to the current 22.2 mpg for light trucks and SUVs).

994. *But cf.* Green Mountain Chrysler Plymouth Dodge Jeep v. Crombie, 508 F. Supp. 2d 295 (D. Vt. 2007) (sustaining Vermont regulations adopting California's greenhouse gas (GHG) emissions standards for new automobiles and finding preemption doctrines did not apply to the interplay between Environmental Protection Agency's (EPA) authority to regulate greenhouse gases from new motor vehicles under Clean Air Act (CAA) and National Highway Traffic Safety Administration's (NHTSA) authority under Energy Policy and Conservation Act (EPCA) to promote energy efficiency by setting mileage standards; and Vermont regulations were not preempted). *See generally* Ann E. Carlson, *Federalism, Preemption, and Greenhouse Gas Emissions*, 27 ENVIRONS ENVTL. L. & POL'Y J. 281 (2003); Steven G. Davison, *Regulation of Emission of Greenhouse Gases and Hazardous Air Pollutants from*

ticipation in international carbon trading by selling unused emissions or sequestration capacity might be found to violate federal dormant commerce clause authority by interfering with foreign relations.[995] Similarly, state regulations that are found to impact out-of state producers or have an extraterritorial effect, through regulatory leakage, might arguably violate the dormant interstate commerce clause.[996] As put forth in Chapter 3 on Food and Agriculture, however, carbon emissions reduction should be seen as a compelling governmental interest and should be immune from commerce clause restrictions unless obvious less restrictive or discriminatory alternatives are readily available. Federal preemption jurisprudence would pose an additional standard to meet as Congressional preemption of the field of foreign affairs is not subject to a compelling state interest exception and state policies that have a direct[997] rather than an incidental[998] or indirect effect are subject to invalidation. The implied tension between state and local initiatives and the failed Kyoto Protocol, however, presents scant evidence of a conflict with federal law or the constitution and preemption is rare and not likely.[999] Carbon taxes would encourage

---

Motor Vehicles, 1 PITT. J. ENVTL PUB. HEALTH L. 1 (2006); J.R. DeShazo & Jody Freeman, *Timing and Form of Federal Regulation: The Case of Climate Change*, 155 U. PA. L. REV. 1499 (2007) (supporting state regulation as it speeds up federal climate control regulation); Sara A. Colangelo, Comment, *The Politics of Preemption: An Application of Preemption Jurisprudence and Policy to California Assembly Bill 1493*, 37 ENVTL. L. 175 (2007). *But see* Nicholle Winters, *Carbon Dioxide: A Pollutant in the Air, But is the EPA Correct that It Is Not an "Air Pollutant"?*, 104 COLUM. L. REV. 1996, 2000–01 (2004) (detailing and comparing the memos, their conclusions, and EPA's decision not to regulate carbon dioxide). *See generally* Michael S. Greve & Jonathan Klick, *Preemption in the Rehnquist Court: A Preliminary Empirical Assessment*, 14 SUP. CT. ECON. REV. 43 (2006).

995. Hannah Chang, *Foreign Affairs Federalism: The Legality of California's Link With the European Union Emissions Trading Scheme*, 37 ENVTL. L. RPTR. 10771 (2007).

996. Erwin Chemerinsky et al., *California, Climate Change, and the Constitution*, 37 ENVTL. L. RPTR. 10653 (2007). *But see* Weisselberg, *supra* note 778 (arguing that the Greenhouse Gas Emissions Performance Standard Act ("SB 1368"), which forbids California utilities from making long-term financial investments or procurement contracts with power plants whose GHG emissions exceed the performance standard, does not violate the Commerce Clause).

997. Zschernig v. Miller, 389 U.S. 429 (1968) (invalidating an Oregon statute that forbade inheritance by nonresident aliens unless his home country offered a reciprocal right of inheritance to U.S. citizens).

998. Clark v. Allen, 331 U.S. 503 (1947) (sustaining California law limiting inheritance by aliens).

999. Robert J. Delahunty, *Federalism Beyond the Water's Edge: State Procurement Sanctions and Foreign Affairs*, 37 STAN. J. INT'L L. 1, 50 (2001). *But see* Chang, *supra* note 995

emission reductions from motor vehicles, building, and energy generation.[1000] Such a carbon tax would be imposed on suppliers of coal, crude oil, and natural gas based upon the carbon content of each.[1001]

There has been a movement to privatize roadways as a means to economize on state services and budget deficits and honor promises and expectations not to raise taxes. According to a survey conducted by the National Association of Realtors and Smart Growth America, Americans are overwhelmingly opposed to private ownership of roads, with 84% opposed to private ownership, and 14% supporting the idea and 66% are opposed to allowing private companies to build, own, and collect tolls for new roads—even if those companies gave a portion of the toll money to the state.[1002]

4. Federal Response

The first and foremost role of the federal government, along with the states, would be full funding for public transport, including a system of high-speed rail between cities within 600 miles or less.[1003] Federal leadership on sustainability in transportation is nonexistent. The current highway statute, the Safe, Accountable, Flexible, Efficient Transportation Equity Act: A Legacy for Users (SAFETEA-LU)[1004] incorporates a vague land use reference that planning needs

---

(presenting the argument that California legislation has a direct effect as it effectively signs California onto the Kyoto Protocol in contradiction of the current official U.S. position).

1000. Burtraw & Portney, in NEW APPROACHES ON ENERGY AND THE ENVIRONMENT, *supra* note 549 at 19 (recommending gradual phasing of increases so as not to destroy the value of existing long-lived investments); Goulder & Nadreau, *in* CLIMATE CHANGE POLICY, *supra* note 342 at 120; Harper, *supra* note11 (advocating carbon taxes as superior to regulation or permit trading). *See also* Soares, *in* EU CLIMATE CHANGE POLICY, *supra* note 549 at 256–73; *Time to Tax Carbon*, L.A. TIMES, May 28, 2007 (editorial endorsing carbon taxes as the most efficient method to reduce emissions as compared to cap-and-trade systems or pure caps); Donehower, *supra* note 116 (arguing that without the incorporation of China and the United States, the world's two largest polluters, the carbon markets may serve as a successful market tool and example of the efficiency of an open market to cost-efficiently solve environmental problems, but will do nothing to curb GHG emissions and limit the effects of climate change).

1001. Goulder & Nadreau, *in* CLIMATE CHANGE POLICY, *supra* note 342 at 131. *See also Time to Tax Carbon*, *supra* note 1000 (editorial endorsing carbon taxes as the most efficient method to reduce emissions as compared to cap-and-trade systems or pure caps).

1002. National Ass'n of Realtors, *supra* note 767.

1003. Bernow, *in* CLIMATE CHANGE POLICY 189, 202–03, *supra* note 93.

1004. Pub. L. No. 109-59, 119 Stat. 1144 (2005).

to consider projects and strategies that will "protect and enhance the environment, promote energy conservation, improve the quality of life, and promote consistency between transportation improvements and State and local planned growth and economic development patterns."[1005] The law does not, however, include a requirement for considering possible land development impacts of planned transportation facilities, that planners should investigate ways to use land use planning and development to reduce travel or to encourage increased transit ridership, and thus contains no lever that would encourage or require the development of an access-efficient transportation planning process.[1006] The focus of transportation planning must shift from simply mobility to include accessibility to work and other necessary destinations. Transportation planning requirements should encourage, reward, and perhaps require, an access-efficient orientation.[1007] Government needs to decentralize governance so that officials need not use air travel. Congress should meet and deliberate virtually and digitally. Subsidies should be targeted to supporting the sciences dealing with public transportation and the expansion of technology education.

Carbon taxes imposed on suppliers of coal, crude oil, and natural gas would encourage emission reductions from motor vehicles, building, and energy generation.[1008] Beginning in 2008, France will tax carbon-generating motor vehicles a carbon tax at the time of sale of $3,800 and higher for cars with higher emissions, while offering rebates of $1,300 for those purchasing low emission vehicles, and nearly $8,000 for those purchasing electric cars and scrapping older cars.[1009] Such a program carries no costs if the taxes fund the subsidy and would be even more effective as an annual tax. Liberal Democrats in Britain are proposing a tax on both air flights and truck freight to generate funds to

---

1005. 23 U.S.C.A. §134(h)(1)(C) (West 2006).

1006. Keith Bartholomew, *The Machine, the Garden, and the City: Toward an Access-Efficient Transportation Planning System*, 37 ENVTL. L. REP. NEWS & ANALYSIS 10593 (2007).

1007. *Id.*

1008. Burtraw & Portney, in NEW APPROACHES ON ENERGY AND THE ENVIRONMENT, *supra* note 549 at 19 (recommending gradual phasing of increases so as not to destroy the value of existing long-lived investments); Goulder & Nadreau, *in* CLIMATE CHANGE POLICY, *supra* note 342 at 120, 131; Harper, *supra* note 11 (advocating carbon taxes as superior to regulation or permit trading). *See also* Soares, *in* EU CLIMATE CHANGE POLICY, *supra* note 549 at 256–73; *Time to Tax Carbon*, *supra* note 1000 (editorial endorsing carbon taxes as the most efficient method to reduce emissions as compared to cap-and-trade systems or pure caps).

1009. *France Slaps Penalties on Gas-Guzzling Cars*, YahooGreen, Dec. 5, 2007, *available at* http://green.yahoo.com/news/afp/20071205/lf_afp/franceclimatetransporttax.html.

finance railway and transit improvements.[1010] The Corporate Average Fuel Economy or "CAFE" standards established in the Energy Policy and Conservation Act,[1011] should be raised requiring significantly higher gas mileage from automobiles sold with a limited amortization period for previously sold vehicles.[1012] Curiously, the Model T Ford was more fuel efficient than today's automobiles and American cars are at a twenty-year low for fuel efficiency.[1013] Subsidies for fossil fuels and highways should be eliminated in favor of taxes on inefficient and polluting vehicles and incentives for renewable energy dependency. Foreign aid should be directed to assisting with local planning to develop viable communities around public transport. The Environmental Protection Agency may and probably must regulate new automobiles to restrict carbon emissions.[1014]

5. International Response

The Kyoto Protocol calls for reductions of motor vehicle and energy carbon emissions to a set percentage under 1990 emissions.[1015] Carbon taxes would

---

1010. *Lib Dems Plan Air Tax to Aid Rail*, BBC NEWS, Aug. 2, 2007, available at http://newsvote.bbc.co.uk/mpapps/pagetools/print/news.bbc.co.uk/2/hi/uk_news/politics/69 (proposing £10 or $20 per ticket and 11p per mile freight tax or about 20¢ similar to freight schemes used in Germany, Austria, Switzerland, and the Czech Republic).

1011. 49 U.S.C. §§ 32901–19 (2000). *See* Center for Biological Diversity v. National Highway Traffic Safety Admin., 508 F.3d 508 (9th Cir. 2007) (finding NHTSA failed to properly consider the benefits of reducing carbon emissions and improperly computed CAFE fuel economy standards under the 1975 Energy Policy and Conservation Act and failed to carry out its responsibilities under the National Environmental Policy Act and failed to issue standards for certain light trucks).

1012. Bernow, *in* CLIMATE CHANGE POLICY 189, 200, *supra* note 93.

1013. CARSON & VAITHEESWARAN, *supra* note 829 at 18.

1014. Massachusetts v. EPA, 127 S. Ct. 1438 (2007) (Clean Air Act authorizes the EPA to regulate greenhouse gas emissions from new motor vehicles in the event that it forms a "judgment" that such emissions contribute to climate change; and can avoid taking regulatory action with respect to greenhouse gas emissions from new motor vehicles only if it determines that greenhouse gases do not contribute to climate change or if it provides some reasonable explanation as to why it cannot or will not exercise its discretion to determine whether they do).

1015. Goulder & Nadreau, *in* CLIMATE CHANGE POLICY, *supra* note 342 at 120; MONBIOT, *supra* note 11 at 48 (Kyoto called for a 5.2% reduction by 2012). *Cf.* Cinnamon Carlarne, *Climate Change—The New "Superwhale" in the Room: International Whaling and Climate Change Politics—Too Much in Common?*, 80 S. CAL. L. REV. 753 (2007) (arguing that international climate control politics looks too much like the failed whaling efforts).

encourage emission reductions from motor vehicles, building construction, and energy generation.[1016] Such a carbon tax would be imposed on suppliers of coal, crude oil, and natural gas based upon the carbon content of each.[1017] Were such a tax imposed internationally, the scheme would eliminate the need for a system of trading carbon permits.[1018] A carbon tax may also be imposed at the state level.

The Kyoto Protocol also contains a call for a Clean Development Mechanism whereby developed countries may receive a carbon reduction credit for financing investments reducing carbon emissions in developing countries and developing countries may purchase carbon permits at a regulated price.[1019] The inclusion of all nations will reduce "carbon leakage" which describes a lowered price of the carbon generating fuel source. As more adopt carbon reduction alternatives, the lowered price from excess supply can result in increased consumption.[1020] Taxing and permitting face additional problems such as how to measure the reduction of methane emissions as a credit to carbon emissions rules.[1021] A controversial topic is how to evaluate carbon trading credits from sequestration initiatives, as carbon reduction obligations would be softened by the trading of credits.[1022] Biotic sequestration includes forestation, wetlands and park protection and expansion, organic roofs, and building systems that exchange carbon dioxide for oxygen.[1023] Morally, industrialized and developed nations should not be able to buy their way out of carbon emissions standards.[1024] A carbon tax would likely be simpler and more effective than a permit trading system.[1025]

1016. Burtraw & Portney, in NEW APPROACHES ON ENERGY AND THE ENVIRONMENT, *supra* note 549 at 19 (recommending gradual pasing of increases so as not to destroy the value of existing long-lived investments); Goulder & Nadreau, *in* CLIMATE CHANGE POLICY, *supra* note 342 at 120; Harper, *supra* note 11 (advocating carbon taxes as superior to regulation or permit trading). *See also* Soares, *in* EU CLIMATE CHANGE POLICY, *supra* note 549 at 256–73; *Time to Tax Carbon*, *supra* note 1000 (editorial endorsing carbon taxes as the most efficient method to reduce emissions as compared to cap-and-trade systems or pure caps).

1017. Goulder & Nadreau, *in* CLIMATE CHANGE POLICY, *supra* note 342 at 131.

1018. *Id* (projecting that a tax of $50 to $150 per ton of carbon, an amount necessary to achieve a 7% reduction would cause coal to double in price, oil and gas increasing by 35–40%, and gasoline rising by 12–14%).

1019. *Id* at 137–39.

1020. *Id* at 140.

1021. *Id* at 142.

1022. *Id* at 142–43.

1023. Bernow, *in* CLIMATE CHANGE POLICY 189, 205–06, *supra* note 93.

1024. Goulder & Nadreau, *in* CLIMATE CHANGE POLICY, *supra* note 342 at 141.

1025. Harper, *supra* note 11 (advocating carbon taxes as superior to regulation or per-

Just as nations can negotiate and agree on strategies, corporations may also establish standards through private treaties and agreements.[1026] Ironically, businesses and consumers will find they receive a significant savings in energy-related costs outweighing the cost of compliance with the Kyoto Protocol or more aggressive and realistic standards.[1027]

International agreements must support transit-oriented development and a reduction of automobile sales, use, and ownership. The expansion of more efficient and affordable public transport should be a target of international agreements and foreign aid.

---

mit trading); Jonathan Baert Wiener, *Designing Global Climate Regulation, in* CLIMATE CHANGE POLICY: A SURVEY 151, 166 (Stephen H. Schneider, Armin Rosencranz & John O. Niles eds. 2002); *Time to Tax Carbon, supra* note 1000 (editorial endorsing carbon taxes as the most efficient method to reduce emissions as compared to cap-and-trade systems or pure caps).

1026. Wiener, *Designing Global Climate Regulation, in* CLIMATE CHANGE POLICY, *supra* note 1025 at ch. 5.

1027. Bernow, *in* CLIMATE CHANGE POLICY 189, 213, *supra* note 93.

Chapter 18

Water Management

It takes about a thousand tons of water to produce one ton of grain that, fed to cows, produces only 18 pounds of meat. Mankind is using 160 billion tons more water each year than is being replenished by rain and fed back into water storages. If this water were carried in water trucks, it would require a 300,000-mile-long convoy of trucks every day—a convoy length 37 times the diameter of the Earth. This is how much water we are using and not replenishing.
JAMES MARTIN, THE MEANING OF THE 21ST CENTURY 4 (2006)

Water is the first great threat from global warming. Just a slight average climate change will threaten those communities already victims of flooding, but will also threaten those living on the coastlines because of water levels that are likely to rise. Even more of a threat is diminishing ice, glaciers, snow pack and over-pumping from aquifers that are reducing the supply of fresh water.[1028] Melting snow packs will also threaten flooding from water that in the past was frozen.[1029] By the end of the century, rising temperatures will demand up to 13% more water for crop irrigation in California.[1030] Most immediately injured will be hundreds of millions in Africa and Asia that have no alternative water supply. Sadly, 90% of the planet's fresh water found in lakes and reservoirs is polluted.[1031] The annual depletion of aquifers worldwide amounts to at least 160 billion tons of water annually.[1032] The planet faces a great challenge in provid-

---

1028. BROWN, *supra* note 36 at 41–56. *See generally* Jon Gertner, *The Future is Drying Up*, N.Y. TIMES, Oct. 21, 2007.
1029. Wagner, *supra* note 1 at ns 24–26 and accompanying text; Matthew D. Zinn, *Adapting to Climate Change: Environmental Law in a Warmer World*, 34 ECOLOGY L.Q. 61, 70–71 (2007).
1030. Zinn, *supra* note 1029 at 69.
1031. WARD, WATER WARS, *supra* note 490 at 205.
1032. MARTIN, THE MEANING OF THE 21ST CENTURY, *supra* note 32 at 67.

ing adequate water to the burgeoning world population.[1033] Even in the United States without global warming, growing population and excessive water consumption[1034] and water pollution[1035] threaten the quality of life. In addition, climate change will also affect water consumption. For example, urban heat islands established in metropolitan areas from hard surfaces, roofs and deforestation significantly increase household water consumption.[1036]

Worldwide, 70% of all water is used for irrigation.[1037] Rice that yields 4 tons per acre uses but a little more water than rice that yields 2 tons an acre; more efficient crops should be substituted for those that demand more water.[1038] Wheat typically produces 50% more calories per unit of water than rice and most protein crops can be modified to produce a high yield with less water.[1039] Using modern farming, 25 gallons of water is used to produce a pound of wheat, but over 5,000 gallons of water are used to produce a pound of beef.[1040] Nonetheless, world meat consumption climbed from 44 million tons in 1950 to 217 million tons in 1999 and is growing.[1041]

Replacing rotating sprinklers with drip irrigation saves a large volume of water and electronic watering systems can reduce water use by 70%.[1042] Hydroponic gardening can produce 18 times the yield per acre of a conventional farm and uses very small amounts of water, but does require careful management and monitoring.[1043] Not only can the seas feed us, but their water is a very precious commodity and must be preserved along with its sources if we

---

1033. THE MULTI-GOVERNANCE OF WATER: FOUR CASE STUDIES (Matthias Finger et al., eds. 2005); WATER IN CRISIS: A GUIDE TO THE WORLD'S FRESH WATER RESOURCES (H. Peter Gleich ed. 1993).

1034. COMMITTEE ON WATERSHED MANAGEMENT, WATER SCIENCE AND TECHNOLOGY BOARD, COMMISSION ON GEOSCIENCES, ENVIRONMENT, AND RESOURCES, NATIONAL RESEARCH COUNCIL, NEW STRATEGIES FOR AMERICA'S WATERSHEDS (1999); SANDRA POSTEL, LAST OASIS: FACING WATER SCARCITY (1992); MARC REISNER, CADILLAC DESERT: THE AMERICAN WEST AND ITS DISAPPEARING WATER (1993); STRATEGIC PLANNING OF SUSTAINABLE URBAN WATER MANAGEMENT (Per-Arne Malmqvist et al., eds. 2006).

1035. THE POISONED WELL: NEW STRATEGIES FOR GROUNDWATER PROTECTION (Eric P. Jorgensen ed. 1989).

1036. Subhrajit Guhathakurta & Patricia Gober, *The Impact of the Phoenix Urban Heat Island on Residential Water Use*, 73 J. AM. PLANNING ASS'N 317 (2007).

1037. MARTIN, THE MEANING OF THE 21ST CENTURY, *supra* note 32 at 70.
1038. *Id.*
1039. *Id.*
1040. *Id* at 74.
1041. *Id.*
1042. *Id* at 70.
1043. *Id* at 75–77.

are to survive. The obvious response in addition to reducing carbon emissions would be to use water more efficiently. There should be no water subsidies as there should be no energy subsidies except to encourage conservation, renewable energy, and more efficient use.[1044]

Additional taxes on water use will encourage conservation in irrigation, lawn care, and waste from excessive personal use. In agriculture, significant improvement in water efficiency would come from a shift from flood or furrow irrigation to a system utilizing overhead sprinklers or, even better, a drip irrigation system.[1045] Conservation would be served by raising the price of water.[1046] Each day we drink 4 to 5 quarts of water in beverages but to produce the food we consume each day takes 2,000 quarts of water.[1047] Even a modest reduction of the consumption of meat, eggs, cheese, and milk would dramatically reduce the use of water. The reduction of such consumption by 100 kilograms per person annually would cut grain use by 30 tons and irrigation water use by 30 billion tons.[1048] It takes a thousand tons of water to produce 1 ton of grain; thus, the annual depleted water from aquifers represents the equivalent loss of 160 million tons of grain.[1049] China has decreased its grain yield by 70 million tons since 1998 due to declining water resources requiring that grain be purchased on the world market which has little grain to sell.[1050] Cattle in feedlots require 7 kilograms of grain to produce a 1 kilogram gain in live weight; pork requires 4 kilograms, poultry just over 2 and fish less than 2 to generate the same gain.[1051] A meal of a hamburger, fries, and a soda from a drive-through restaurant requires 1,500 gallons of water when growing the potatoes, grain for the bun and feed for the cow, and water for the soda.[1052]

A shift from coal-fired power plants that use vast amounts of water for thermal cooling to wind- or solar-generated power would save considerable water use.[1053] Desalination may increase in affordability but it is energy intensive in pumping water through the plant and in pumping water into aquifers and then pumping water out and deposits brine which threatens the environment.[1054] Rel-

---

1044. BROWN, *supra* note 36 at 167.
1045. *Id* at 168.
1046. *Id* at 169.
1047. MARTIN, THE MEANING OF THE 21ST CENTURY, *supra* note 32 at 66.
1048. BROWN, *supra* note 36 at 170.
1049. MARTIN, THE MEANING OF THE 21ST CENTURY, *supra* note 32 at 67.
1050. *Id* at 69.
1051. BROWN, *supra* note 36 at 171.
1052. ROGERS & KOSTIGEN, *supra* note 209 at 15.
1053. BROWN, *supra* note 36 at 170.
1054. Zinn, *supra* note 1029 at 69–70; Gertner, *supra* note 1028.

atively slight investments in industry and domestic use can preserve water and dramatically reduce costs through recycling water.[1055]

Waste management that has been based on water must also be changed to conserve on water. Using water toilets and vast sewage systems is extraordinarily wasteful of water and can be replaced by compost toilets and water-efficient appliances and showerheads, dishwashers, clothes washers, and water recycling.[1056]

The use of the seas to mitigate global climate change is problematic. Wave-generated power holds some promise but raises environmental concerns. Subseabed sequestration of carbon is a potential strategy to avert the full force of global warming yet it raises many international legal issues as to whether it might be characterized as dumping in violation of statutes and treaties protecting the seas.[1057] In addition to conservation, the future is in desalination of sea water for drinking water.[1058]

1. Local Response

At the local community level, water must be conserved through sustainable behavior both personally and with respect to development. Water-reduced appliances and water-saver or organic toilets and the adoption of water-saving habits, can help reduce demand. In wet climates, water can be collected in cisterns and used for toilet flushing. We can bath less frequently and without as much water. Through taxation and regulation the development of swimming pools and hot tubs can be discouraged. Treating sewage and purifying surface and waste water locally is preferable rather than to pipe and pump it to huge treatment plants.

Communities and water districts can establish water pricing that encourages conservation.[1059] Adequate pricing, marketing policies in transferring water rights to third parties, and water-use efficiency improvements are major tools for water-scarce regions.[1060]

1055. WARD, WATER WARS, *supra* note 490 at 203–04.
1056. BROWN, *supra* note 36 at 216–19.
1057. Weeks, *supra* note 427.
1058. Goodell, *The Prophet of Climate Change, supra* note 70. *See also Reverse Osmosis and Toray Membrane: Can Desalination Finally Solve Water Scarcity?*, VERDEXCHANGE, Nov. 12, 2007, *available at* http://verdexchange.org/200710/reverse.html.
1059. Turman, *in* CLIMATE CHANGE POLICY, *supra* note 44 at 106, *citing* 2 CALIFORNIA ENERGY COMMISSION, *supra* note 806; FIELD ET AL., *supra* note 806; KNOX & FOLEY, *supra* note 806; SMITH & TIRPAK, *supra* note 806.
1060. Frederick & Gleick, *in* CLIMATE CHANGE, *supra* note 48 at 76–77.

In dry climates, lawns may be prohibited in favor of local indigenous plants, succulents, sedums, which are slow growing, drought tolerant plants, and other dry landscaping that can be adequately watered through precipitation. In dry climates, cities can offer programs to replace grass lawns with desert plants. Landowners could be required to collect rain and surface water runoff and use it for irrigation. Water management should be a part of the spatial planning process and comprehensive plans should include a water management element.[1061] Water should play a key role in assessing the carrying capacity of land when planning and regulating for growth.[1062] Water policy may be a superior method of regulating and controlling growth and sprawl.[1063]

2. Regional Response

At the regional level, water supplies can be assessed and regional needs and supply allocated. Development should not occur absent an adequate water supply.[1064]

3. State Response

States will have to preserve scarce water supplies and restrict development according to availability and anticipated shortages. More investment must be made in water supply preservation and research into farming and irrigation so as to preserve sufficient drinking water. Water-intensive crops, such as cotton, ethanol and meat-destined grains should not be grown. States must as-

---

1061. Johan Woltjer & Niels Al, *Integrating Water Management and Spatial Planning*, 73 J. Am. Planning Ass'n 211 (2007).

1062. A. Dan Tarlock & Sarah B. Van. De Wetering, *Water and Western Growth*, 59 Planning & Envt'l L. 3 (No. 5 May 2007) (discussing limiting growth through water management and land use planning around water planning).

1063. Wet Growth: Should Water Law Control Land Use? (Craig Anthony (Tony) Arnold ed. 2005); Craig Anthony (Tony) Arnold, *Is Wet Growth Smarter Than Smart Growth?: The Fragmentation and Integration of Land Use and Water*, 35 Env't L. Rptr. 10152 (Mar. 2005).

1064. Cal. Gov. Code § 66473.7(a)(2) (2003) (requiring conditioning certain residential developments on the availability of an adequate water supply); Ryan Waterman, Comment, *Addressing California's Uncertain Water Future by Coordinating Long-Term Land Use and Water Planning: Is a Water Element in the General Plan the Next Step?*, 31 Ecology L.Q. 117 (2004).

sure that water is allocated fairly and priced according to its value. Local government must be restricted in allowing sprawl development to accommodate population projections. States can mandate that water management should be a part of the spatial planning process and comprehensive plans should include a water management element.[1065] Water should play a key role in assessing carrying capacity of land and in planning and regulating for growth.[1066]

4. Federal Response

The federal government must consider the supply of potable water and how development should be directed to assure a reasonable supply. Immigration should be limited to the capacity of infrastructure and the adequacy of water.

Water conservation can be broadly addressed by decisions about regulation and spending available for reforestation and elimination of the most water-intensive crop farming. It is essential that taxes be imposed on meat due to the extraordinary amount of water used to produce it, particularly in the growing of feed. Foreign aid should be targeted at protecting water sources and the fair allocation of water supplies. Should desalination plants become a feasible water supplier, this technology would have to be produced and shared with those facing the worst water shortages. Water management also includes protecting against flooding. The Army Corps of Engineers too often constructs pork barrel projects for powerful politicians and some projects have been considered economically indefensible and environmentally destructive.[1067] The Corps has been charged as the primary cause of the destruction of New Orleans following Hurricane Katrina.[1068] The federal government should condition federal

---

1065. Woltjer & Al, *supra* note 1061.
1066. Tarlock & Van. De Wetering, *supra* note 1062 (discussing limiting growth through water management and land use planning around water planning).
1067. Grunwald, *Setting the Stage for More Katrinas*, *supra* note 413. See also Jennifer Cutraro, *Engineering Change: A Brief History of the Creation and Growth of the Army Corps*, Grist, Mar. 17, 2008, *available at* http://www.grist.org/feature/2008/03/17/cutraro/index.html?source=rss; Emily Gertz, *Tempting Fate: Fifteen Years After the Great Flood of 1993, Floodplain Development is Booming*, Grist, Mar. 19, 2008, *available at* http://www.grist.org/feature/2008/03/19/gertz/; Michael Grunwald, *Cry Me a River*, Grist, Mar. 18, 2008, *available at* http://www.grist.org/feature/2008/03/18/grunwald/index.html.
1068. *In re* Katrina Canal Breaches Consolidated Litigation, 471 F. Supp. 2d 684 (E.D. La. 2007) (rejecting defendants' motion to dismiss and rejecting immunity under Flood Control Act and Federal Tort Claims Act); Grunwald, *Setting the Stage for More Katrinas*, *supra* note 413.

infrastructure grants and investments on state land and water planning including balancing supply and population, reducing and removing water pollution, storm water treatment, and conservation.[1069] Water is often an interstate problem calling for national regulation to assure fair allocation and sustainability.

5. International Response

International cooperation is necessary to protect the nearly 80% of the planet that is facing flooding and water shortages. Where no feasible alternative is available, some may need to be relocated from islands and coastal areas. In addition, development should only be encouraged where there is adequate water. Conservation should be undertaken everywhere and emergency water supplies may have to be delivered, despite fuel use and carbon emissions, where there are droughts.

1069. Bruce Babbitt, *The Case for Conditioning Federal Infrastructure Investment on State Land and Water Planning*, 73 J. AM. PLANNING ASS'N 146 (2007).

Chapter 19

Conclusion

The really inconvenient truth, which no legislator or former legislator will publicly acknowledge, is that to even attempt to reconcile the American way of life with the sustainability of the planet will require decisive action and dramatic change. At every turn both state and federal legislators—even those with the best intentions—will seek to avoid environmental measures which might interfere with the relative luxury of heating or cooling your homes or driving or flying whenever and however you wish, and substitute measures, like biofuels, which transfer the cost onto less powerful people.

     George Monbiot, Heat: How to Stop the
     Planet From Burning viii (2007)

We have at most ten years—not ten years to decide upon action, but ten years to alter fundamentally the trajectory of global greenhouse emissions.... Such an outcome is still feasible in the case of global warming, but just barely.

     James Hansen, *The Threat to the Planet*, New York
     Review of Books at 53 (No. 12, July 13, 2006)

I would sooner expect a goat to succeed as a gardener than expect humans to become stewards of the Earth.

     James Lovelock, *quoted in* Jeff Goodell,
     The Prophet of Climate Change: James Lovelock,
     Rolling Stone.com, Oct. 17, 2007, available at
     http://www.rollingstone.com/politics/story/16956300/
     the_prophet_of_climate_change_james_lovelock

How could I look my grandchildren in the eye and say I knew about this and I did nothing?

     Sir David Attenborough, cited in Paul Brown, Global
     Warning: The Last Chance for Change 6 (2007)

New Orleans, Sept. 6, 2005. Source: FEMA/Jocelyn Augustino.

[W]orld leaders must make courageous decisions to save the planet before it is too late.

Pope Benedict XVI[1070]

The risk is too grave not to launch an aggressive fight for the planet and survival of the human race. Technology, research, and development will generate opportunities, improvements, and adjustments, but the nation possesses the capital, intelligence, and tools to reduce emissions and design liveable, more pleasant communities with the dividend of living sustainably. It is feasible that a worldwide annual expenditure of $161 billion could save the planet, avoid extinction, and solve many of the intractable problems from hunger and poverty to education, environmental sustainability, and an improved quality of life.[1071]

1070. Philip Pullella, *Save the Planet Before it's Too Late, Pope Urges*, YAHOO NEWS, Sept. 2, 2007, *available at* http://news.yahoo.com/s/nm/20070902/wl_nm/pope_dc_1;_ylt=Aq3_KwhkpAXqiTBnDtyJ67BkM3wV.

1071. BROWN, *supra* note 36 at 255–265.

Marble Road—excessive urban sprawl, deforestation, and erosion led to unsustainability, loss of sea access, and destruction of the ancient world's fourth largest city, Ephesus, Turkey. Source: James A. Kushner.

Afterword

As an academic, one has the freedom to inquire into virtually any subject, often receiving research grants to encourage the undertaking. Many authors and writers must consider market demand, sales potential, or placement prestige. Luckily for me, there exists a broad array of specialty publishing houses and academic journals that will publish something of interest. It is in that environment that I have been writing for the past thirty-five years.

I have always used my senses, emotions, and an insatiable love of reading books from a wide array of disciplines to consider the world around me, educate myself and to determine the most interesting or compelling topics to me. Ever since the 1960s, when I was a Peace Corps Volunteer, a poverty volunteer in VISTA, (our domestic Peace Corps), and a legal services attorney representing the poor, I have focused on issues of social equity, solutions for poverty, the politically powerless, and the quality of life.

My first book was *Apartheid in America* (1980). It explored the origins, continuing causes, and potential mitigation policies for racial segregation in the United States which, in the late 1970s, I believed to be the central cause of discrimination, suffering, and the decline of the quality of American life. My second book was *Housing and Community Development* (1981), a collaboration with some of the top housing law professors and theorists in the nation, focusing on poverty and housing, affordable housing, regeneration of declining cities and neighborhoods, and housing justice. *Housing and Community Development*, which is a textbook heading for its fourth edition and written for law, planning and urban studies students, reflected the view that inadequate housing, segregation, and declining cities were at the center of problems in the United States and that they must be resolved for the nation that aspired to be the greatest nation to reach that goal.

My next book, *Fair Housing: Discrimination in Real Estate, Community Development and Revitalization* (1983), was based on my continued belief that racial segregation and discrimination in housing and land use were at the center of injustice, poverty, and unfairness in America and that enforcement of civil rights laws could provide a strategy to reduce the causes of that injustice. I then wrote *Government Discrimination: Equal Protection Law and Litigation* (1988), now in its 19th edition, which extended my study of discrimination to all facets of bias, private as well as public, and extended the scope of my work to cover all victims of discrimination both governmental and private, and not limited to ethnic minorities.

Subdivision Law and Growth Management (1991), now a two-volume work in its second edition, and its 17th annually updated version, was written after a sabbatical studying urban sprawl. It reflected my sense that urban sprawl, de struction of open space, and the urban design of low density suburbs requiring personal automobiles were making the pursuit of accessibility, environmental sustainability, and liveable communities an impossibility; and that better communities could be generated from limiting low-density growth, rediscovering urban development, and planning for social integration. Along with my coauthor, Dan Selmi, I wrote *Land Use Regulation* (1999), a textbook for law and urban planning students now in its third edition with Ed Ziegler, in which we tie together American urban planning law and the design of communities with issues of housing equity, sustainability, and transportation and linked growth management, sprawl, and smart growth with urban redevelopment, new urbanism, brownfield development, and regeneration of cities and neighborhoods.

In the 1990s, I began extensively traveling, studying, publishing,[1072] and teaching in Europe, comparing urban development, land use policies and practices, to those of the United States.

1072. Some of my published articles during this period included: (1) *Urban Transportation Planning*, 4 URB. L. & POL'Y 161 (1981); (2) *A Tale of Three Cities: Land Development and Planning for Growth in Stockholm, Berlin, and Los Angeles*, 25 URB. LAW. 197 (1993); (3) *Growth Management and the City*, 12 YALE L. & POL'Y REV. 68 (1994); (4) *Growth for the Twenty-First Century: Tales from Bavaria and the Vienna Woods—Comparative Images of Urban Planning in Munich, Salzburg, Vienna, and the United States*, 29 URB. LAW. 911, 924 (1997), *reprinted as modified*, 6 S. CAL. INTERDISC. L.J. 89 (1997); (5) *A Comparative Vision of the Convergence of Ecology, Empowerment, and the Quest for a Just Society*, 52 U. MIAMI L. REV. 931 (1998); (6) *Smart Growth: Urban Growth Management and Land-Use Regulation Law in America*, 32 URB. LAW. 211 (2000); (7) *Social Sustainability: Planning for Growth in Distressed Places—the German Experience in Berlin, Wittenberg, and the Ruhr,"* 3 WASH. U. J. L. & POL'Y 849 (2000), *published in* EVOLVING VOICES IN LAND USE LAW Ch. 13 (Washington University Journal of Law & Policy 2000); (8) *Planning for Downsizing: a Comparison of the Economic Revitalization Initiatives in American Communities Facing Military Base Closure with the German Experience of Relocating the National Capital from Bonn to Berlin*, 33 URB. LAW. 119 (2001); (9) *Smart Growth, New Urbanism, and Diversity: Progressive Planning Movements in America and Their Impact on Poor and Minority Ethnic Populations*, 21 UCLA J. ENVTL L. & POL'Y 45 (2002/2003); (10) *Walt Disney and the Quest for Community*, 19 J. HOUSING & BUILT ENV'T 129 (2004); (11) *Car Free Housing Developments:*

Out of those experiences, I wrote *Comparative Urban Planning Law* (2003). That text book reflected my view that European compact cities were more equitable and sustainable than American urban design and that most nations, particularly those in Latin America, were actually developing more liveable communities despite their relative poverty.

After living and teaching in the Netherlands and realizing that a higher standard of living was available from compact cities linked by public transport and that a pedestrian lifestyle, a design far more sustainable and enjoyable than one involving personal automobiles, I wrote *The Post-Automobile City* (2004). The book sought to inform the public and decisionmakers how the United States could be converted to a more pedestrian-friendly society and the quality of life greatly improved. Being a realist, I could not then imagine how America could give up the automobile completely because it is such a large part of the economy and popular culture, and so sought to imagine a society where cars and pedestrians could co-exist more sustainably.

My next project was to look at health care, the cost of delivery of health services in the United States and the paradox of spending more and more of our GDP on health care while the health of our population was dramatically declining. I then wrote *Healthy Cities—The Intersection of Urban Planning, Law, and Health*, in which I came to recognize that an automobile-based transport system could never be sustainable and could never yield a sustainable and liveable community. Growth management, walkable neighborhoods, convenient public transport, and better planning for health care and a healthful lifestyle was essential to a high quality of life, manageable health, and environmental sustainability.

It was almost automatic that my quest carried me to the conundrum of global climate change, the central and perhaps last challenge of our civilization.

Towards Sustainable Smart Growth and Urban Regeneration Through Car-Free Zoning, Car-Free Redevelopment, Pedestrian Improvement Districts, and New Urbanism, 23 UCLA J. ENVTL. L. & POL'Y 1 (2005); (12) *New Urbanism: Urban Development and Ethnic Integration in Europe and the United States*, 5 U. MD. J. RACE, GENDER & CLASS 27 (2005); (13) *City Life in the Age of High Technology*, 37 URB. LAW. 893 (2005); (14) *Budapest, Istanbul & Warsaw: Institutional and Spatial Change*, 20 J. HOUSING & BUILT ENV'T 201 (2005); (15) *Brownfield Redevelopment Strategies in the United States*, 22 GA. ST. U. L. REV. 857 (2006), and (16) *Urban Planning and the American Family*, 36 STETSON L. REV. 67 (2006).

Table of Authorities

John Abatzoglou et al., *A Primer on Global Climate Change and Its Likely Impacts*, in CLIMATE CHANGE: WHAT IT MEANS FOR US, OUR CHILDREN, AND OUR GRANDCHILDREN 11 (Joseph F.C. DiMento & Pamela Doughman eds. 2007)

Michael Abrams, *Invisible Hitchhikers: Slide Behind the Wheel and your Biggest Health Threat May Not be the Other Drivers On the Road, but the Invisible Toxins Riding Shotgun*, MS.COM, Mar. 15, 2008, *available at* http://health.msn.com/health-topics/asthma/articlepage.aspx?cp-documentid=100185275&GT1=31005

Gregory M. Adams, Comment, *Bringing Green Power to the Public Lands: The Bureau of Land Management's Authority and Discretion to Regulate Wind-Energy Developments*, 21 J. ENVTL. L. & LITIG. 445 (2006)

Richard M. Adams et al., *Impacts on the Agricultural Sector*, in CLIMATE CHANGE: SCIENCE, STRATEGIES, & SOLUTIONS 25, 40 (Eileen Claussen, Vicki Arroyo Cochran & Debra P. Davis eds. 2001)

Matthew D. Adler, *Corrective Justice and Liability for Global Warming*, 155 U. PA. L. REV. 1859 (2007)

Myles Allen et al., *Scientific Challenges in the Attribution of Harm to Human Influence on Climate*, 155 U. PA. L. REV. 1353 (2007)

AMERICAN SOCIETY OF LANDSCAPE ARCHITECTS, THE SUSTAINABLE SITES INITIATIVE—STANDARDS AND GUIDELINES: PRELIMINARY REPORT (2007), *available at* http://greenerbuildings.com/news_detail.cfm?NewsID=36194&print=true

ANI, *Family Planning Rules in China Tightened*, YAHOO NEWS INDIA, Sept. 15, 2007

Craig Anthony (Tony) Arnold, *Is Wet Growth Smarter Than Smart Growth?: The Fragmentation and Integration of Land Use and Water*, 35 ENV'T L. RPTR. 10152 (Mar. 2005)

Arup to Design Infrastructure for Personal Rapid Transport System at Heathrow, Apr. 11, 2005 (press release), *available at* http:www.arup.com/newsitem.cfm?pageid=7565

Associated Press, *Climate Report: 'Highway to Extinction:' Dire Predictions Includes Loss of Species, Increasing Scarcity of Water*, MSNBC.COM, April, 1, 2007

Associated Press, *Climate Report Sound Dire Warnings: Global Warming Could Mean Hundreds of Millions without Water*, MSNBC.COM, Mar. 10, 2007

Associated Press, *Corn Boom Could Expand 'Dead Zone' in Gulf*, MSNBC, Dec. 17, 2007, *available at* http://www.msnbc.msn.com/id/22301669/

Associated Press, *Study: Corn Ethanol No Climate Solution: Greenhouse Emissions Much Higher if Land Use Factored In, Researchers Say*, MSNBC.COM, Feb. 7, 2008, available at http://www.msnbc.msn.com/id/23057867/

Associated Press, *Most Polar Bears Could Die Out by 2050: U.S. Geological Survey Says Two-Thirds Could Vanish Because of Ebbing Ice*, MSNBC, Sept. 8, 2007, *available at* http://www.msnbc.msn.com/id/20645362/

Robert U. Ayers, *The Energy We Overlook*, WORLD-WATCH, Nov.–Dec., 2001, at 30, *cited in* JAMES GUSTAVE SPETH, RED SKY AT MORNING: AMERICA AND THE CRISIS OF THE GLOBAL ENVIRONMENT 64 (2004)

Bruce Babbitt, *The Case for Conditioning Federal Infrastructure Investment on State Land and Water Planning*, 73 J. AM. PLANNING ASS'N 146 (2007)

Chris Backes & Reinske Teuben, *Legal Aspects of the Dutch Approach to CO_2* in CLIMATE CHANGE AND THE KYOTO PROTOCOL: THE ROLE OF INSTITUTIONS AND INSTRUMENTS TO CONTROL GLOBAL CHANGE 128, 129 (Michael Faure et al. eds. 2003)

Zoya E. Bailey, Comment, *The Sink that Sank the Hague: A Comment on the Kyoto Protocol*, 16 TEMP. INT'L & COMP. L.J. 103 (2002)

DAVID BANISTER, UNSUSTAINABLE TRANSPORT: CITY TRANSPORT IN THE NEW CENTURY (2005)

Jonathan Barnett & Kristina Hill, *Design for Rising Sea Levels*, HARVARD DESIGN MAGAZINE (No. 27 Fall2007/Winter 2008), *available at* http://www.gsd.harvard.edu/research/publications/hdm/current/27_BarnettHill.html

Keith Bartholomew, *The Machine, the Garden, and the City: Toward an Access-Efficient Transportation Planning System*, 37 ENVTL. L. REP. NEWS & ANALYSIS 10593 (2007)

CHARLES BARTSCH & ELIZABETH COLLATON, BROWNFIELDS: CLEANING AND REUSING CONTAMINATED PROPERTIES (1997)

Keith Bea et al., Federal Emergency Management Policy Changes After Hurricane Katrina: A Summary of Statutory Provisions (Dec. 15, 2006), available at http://www.fas.org/sgp/crs/homesec/RL33729.pdf

TIMOTHY BEATLEY, GREEN URBANISM: LEARNING FROM EUROPEAN CITIES (2000)

Timothy Beatley & Richard Collins, *Smart Growth and Beyond: Transitioning to a Sustainable Society*, 19 Va. Envtl. L.J. 287 (2000)

Timothy Beatley & Kristy Manning, The Ecology of Place: Planning for Environment, Economy, and Community (1997)

Sharon Begley, *The Truth About Denial,.* Newsweek, Aug. 13, 2007, available at http://www.msnbc.msn.com/id/20122975/site/newsweek/page/0/

Marisol Bello, *Cities Cultivate 2 Types of Green*, USA Today, Dec. 12, 2007, available at http://www.usatoday.com/news/nation/environment/2007-12-12-green-jobs_N.htm

John J. Berger, *Renewable Energy Sources as a Response to Global Climate Concerns*, in Climate Change Policy: A Survey 411, 440 (Stephen H. Schneider, Armin Rosencranz & John O. Niles eds. 2002)

Steven Bernow et al., *Carbon Abatement with Economic Growth: A National Strategy*, in Climate Change Policy: A Survey 189 (Stephen H. Schneider, Armin Rosencranz & John O. Niles eds. 2002)

Sharon Bernstein & Francisco Vara-Orta, *Near the Rails but Still on the Road*, L.A. Times, Jun. 30, 2007

Daniel Bianchi, *Cross Compliance: The New Frontier in Granting Subsidies to the Agricultural Sector in the European Union*, 19 Geo. Int'l Envtl. L. Rev. 817 (2007)

Edwin Black, Internal Combustion: How Corporations and Governments Addicted the World and Derailed the Alternatives (2006)

Erik B. Bluemel, *Unraveling the Global Warming Regime Complex: Competitive Entropy in the Regulation of the Global Public Good*, 155 U. Pa. L. Rev. 1981 (2007)

Christoph Böhringer & Michael Finus, *The Kyoto Protocol: Success or Failure?*, in Climate-Change Policy 254 (Dieter Helm ed.2005)

Steve Bonta, *The United States Should Not Increase Foreign Aid to Developing Nations*, in Developing Nations 110 (Berna Miller & James D. Torr eds. 2003)

Jan-Tjeerd Boom & Andries Nentjes, *Alternative Design Options for Emissions Trading: A Survey and Assessment of the Literature*, in Climate Change and the Kyoto Protocol: The Role of Institutions and Instruments to Control Global Change 45 (Michael Faure et al. eds. 2003)

James Boyce, *The Greening of Politics: Seven Years of Rapid Change*, msn.com, Sept. 1, 2007, available at http://stopglobalwarming.msn.com/article.aspx?&cp-documentid=5288548.

Gary Braasch, Earth Under Fire: How Global Warming is Changing the World 7–8 (2007)

Keith Bradsher & David Barboza, *Pollution From Chinese Coal Casts a Global Shadow*, N.Y. Times, Jun. 11, 2006

Donald Brown, American Heat: Ethical Problems with the United States Response to Global Warming (2002)

Lester R. Brown, Plan B 2.0: Rescuing a Planet Under Stress and a Civilization in Trouble (2006)

Paul Brown, Global Warning: The Last Chance for Change (2007)

Paul Brown, *Ice Caps Melting Fast: Say Goodbye to the Big Apple?*, AlterNet, Oct. 15, 2007, available at http://www.alternet.org/story/64735/

Peter Buchanan, Ten Shades of Green: Architecture and the Natural World (2005)

Building Without Borders: Sustainable Construction for the Global Village (Joseph F. Kennedy ed. 2004)

Harriet Bulkeley & Michele M. Bewtsill, Cities and Climate Change: Urban Sustainability and Global Environmental Governance (2003)

Thomas G. Burns, *Global Climate Change: A Business Perspective, in* Climate Change Policy: A Survey 279 (Stephen H. Schneider, Armin Rosencranz & John O. Niles eds. 2002)

Dallas Burtraw & Paul R. Portney, *A Carbon Tax to Reduce the Deficit*, in New Approaches on Energy and the Environment: Policy Advice for the President 19 (Richard D. Morgenstern & Paul R. Portney eds. 2004)

Jason R. Busch et al., *Tax and Financial Incentives for Green Building*, 30 Los Angeles Law. 15 (Jan. 2008)

President George W. Bush, State of the Union Address (Jan. 23, 2007), *available at* 1961.1http://www.whitehouse.gov/news/releases/2007/01/20070123-2.html

2 California Energy Commission, Global Climate Change Potential Impacts & Policy Recommendations (1991), *cited in* Eleanor G. Turman, *Regional Impact Assessments: A Case Study of California, in* Climate Change Policy: A Survey 89 (Stephen H. Schneider, Armin Rosencranz & John O. Niles eds. 2002)

Alan Carlin, *Global Climate Change Control: Is There a Better Strategy Than Reducing Greenhouse Gas Emissions?*, 155 U. Pa. L. Rev. 1401 (2007)

Cinnamon Carlarne, *Climate Change—The New "Superwhale" in the Room: International Whaling and Climate Change Politics—Too Much in Common?*, 80 S. Cal. L. Rev. 753 (2007)

Ann E. Carlson, *Federalism, Preemption, and Greenhouse Gas Emissions*, 27 ENVIRONS ENVTL. L. & POL'Y J. 281 (2003)

Ann Carlson & Tim Malloy, *Special Edition: California's AB 1493: Trendsetting or Setting Ourselves Up To Fail?*, 21 UCLA J. ENVTL. L. & POL'Y 97 (2003)

Christopher Carr & Flavia Rosembuj, *Flexible Mechanisms for Climate Change Compliance: Emission Offset Purchases Under the Clean Development Mechanism*, 16 N.Y.U. ENVTL. L.J. 44 (2008)

IAIN CARSON & VIJAY V. VAITHEESWARAN, ZOOM: THE GLOBAL RACE TO FUEL THE CAR OF THE FUTURE (2007)

RACHEL CARSON, SILENT SPRING (1962)

ROBERT CERVERO, THE TRANSIT METROPOLIS: A GLOBAL INQUIRY (1998)

Anne Marie Chaker, *Clothesline Has Neighbors Bent Out of Shape in Bend*, REAL ESTATE JOURNAL.COM, Sept. 24, 2007

Lisa Chamberlain, *Skyfarming*, N.Y. MAGAZINE, Apr. 9, 2007, *available at* http://www.nymag.com/news/features/30020

Hannah Chang, *Foreign Affairs Federalism: The Legality of California's Link With the European Union Emissions Trading Scheme*, 37 ENVTL. L. RPTR. 10771 (2007)

David S. Chapman & Michael G. Davis, *Global Warming—More than Hot Air?*, 27 J. LAND RESOURCES & ENVTL. L. 59 (2007)

Erwin Chemerinsky *et al.*, *California, Climate Change, and the Constitution*, 37 ENVTL. L. RPTR. 10653 (2007)

Cheney on Global Warming: Vice President's Views at Odds with Majority of Climate Scientists, ABC NEWS TECHNOLOGY AND SCIENCE, Feb. 23, 2007

Jia-Rui Chong, *Global Warming: Enough to Make You Sick*, LOS ANGELES TIMES, Feb. 25, 2007

CLIMATE CHANGE AND THE KYOTO PROTOCOL: THE ROLE OF INSTITUTIONS AND INSTRUMENTS TO CONTROL GLOBAL CHANGE (Michael Faure *et al.* eds. 2003)

CLIMATE CHANGE: SCIENCE, STRATEGIES, & SOLUTIONS 63 (Eileen Claussen, Vicki Arroyo Cochran & Debra P. Davis eds. 2001)

CLIMATE CHANGE POLICY: A SURVEY (Stephen H. Schneider, Armin Rosencranz & John O. Niles eds. 2002)

CLIMATE CHANGE: WHAT IT MEANS FOR US, OUR CHILDREN, AND OUR GRANDCHILDREN (Joseph F.C. DiMento & Pamela Doughman eds. 2007)

The Coal Trap, N.Y. TIMES, May 30, 2007

Casey Cohn, Student Article, *The Brownfields Revitalization and Environmental Restoration Act: Landmark Reform or a "Trap for the Unwary"?*, 12 N.Y.U. ENVTL. L.J. 672 (2004)

Sara A. Colangelo, Comment, *The Politics of Preemption: An Application of Preemption Jurisprudence and Policy to California Assembly Bill 1493*, 37 ENVTL. L. 175 (2007)

COMMITTEE ON WATERSHED MANAGEMENT, WATER SCIENCE AND TECHNOLOGY BOARD, COMMISSION ON GEOSCIENCES, ENVIRONMENT, AND RESOURCES, NATIONAL RESEARCH COUNCIL, NEW STRATEGIES FOR AMERICA'S WATERSHEDS (1999)

Jeffrey S. Conner, *Managing the Risks of LEED Certification*, 30 LOS ANGELES LAW. 10 (Jan. 2008)

Amy Cortese, *As the Earth Warms, Will Companies Pay?*, N.Y. TIMES, Aug. 18, 2002, §3, at 6

ANN COULTER, GODLESS (2006)

Jennifer Couzin, *Living in the Danger Zone*, 319 SCIENCE 748 (Feb. 2008)

Stan Cox, *SUVs Without Wheels*, COMMON DREAMS.NEWS CENTER, Mar. 13, 2008, *available at* http://www.commondreams.org/archive/2008/03/13/7650/

J. H. CRAWFORD, CARFREE CITIES (2000)

MICHAEL CRICHTON, STATE OF FEAR (2004)

Jennifer Cutraro, *Engineering Change: A Brief History of the Creation and Growth of the Army Corps*, GRIST, Mar. 17, 2008, *available at* http://www.grist.org/feature/2008/03/17/cutraro/index.html?source=rss

Jocelyn D'Ambrosio, Student Article, *Alternative Fuels: An Evaluation of Corn Ethanol, Cellulosic Ethanol, and Gasoline*, 37 ENVTL. L. REP. NEWS & ANALYSIS 10615 (2007)

ANDREW E. DASSLER & EDWARD A. PARSON, THE SCIENCE AND POLITICS OF GLOBAL CLIMATE CHANGE: A GUIDE TO THE DEBATE (2006).

Mark K. Dausch, Comment, *Analyzing a Municipality's Authority to Enact the Model Ordinance for Wind Energy Facilities in Pennsylvania*, 45 DUQ. L. REV. 47 (2006)

Steven G. Davison, *Regulation of Emission of Greenhouse Gases and Hazardous Air Pollutants from Motor Vehicles*, 1 PITT. J. ENVTL PUB. HEALTH L. 1 (2006)

Robert J. Delahunty, *Federalism Beyond the Water's Edge: State Procurement Sanctions and Foreign Affairs*, 37 STAN. J. INT'L L. 1 (2001)

Stephen T. Del Percio, Student Article, *The Skyscraper, Green Design & LEED Green Building Rating System: The Creation of Uniform Sustainable Standards for the 21st Century or the Perpetuation of an Architectural Fiction?*, 28 ENVIRONS 117 (2004)

Delucchi Study Finds that U.S. Motorists Do Not Pay Their Way, STREETSBLOG, Sept. 20th, 2007

Ralph A. DeMeo et al., *Insuring Against Environmental Unknowns*, 23 J. LAND USE & ENVTL. L. 61 (2007)

Judy Dempsey, *Coalition Tackles Germany's Falling Birth Rate: Financial Package for Working Women Wins Wide Support*, INT'L HERALD TRIBUNE, Jun. 15, 2006

DEPARTMENT OF HOMELAND SECURITY, NATIONAL RESPONSE FRAMEWORK (Jan. 2008)

John C. Dernbach, *Overcoming the Behavioral Impetus for Greater U.S. Energy Consumption*, 20 GLOBAL BUS. & DEV. L. J. 15 (2007)

John C. Dernbach, *Sustainable Development as a Framework for National Governance*, 49 CASE W. RESERVE L. REV. 1 (1998)

John C. Dernbach, *U.S. Policy*, in GLOBAL CLIMATE CHANGE AND U.S. LAW 61 (Michael B. Gerrard ed. 2007)

J.R. DeShazo & Jody Freeman, *Timing and Form of Federal Regulation: The Case of Climate Change*, 155 U. PA. L. REV. 1499 (2007)

Jonathan Donehower, Comment, *Analyzing Carbon Emissions Trading: A Potential Cost Efficient Mechanism to Reduce Carbon Emissions*, 38 ENVTL. L. 177 (2008)

Elizabeth Douglass, *His Passion for Solar Still Burns*, L.A. TIMES, Nov. 10, 2007

Wybe Th. Douma, *The European Union, Russia and the Kyoto Protocol*, in EU Climate Change Policy: The Challenge of New Regulatory Initiatives 51 (Marjan Peeters & Kurt Deketelaere eds. 2006)

Dave Downey, *Smart Growth Policy Revision Needed, Some Say*, NORTH COUNTY TIMES, Dec. 6, 2007, *available at* http://www.nctimes.com/articles/2007/12/04/news/top_stories/1_04_1512_3_07.txt

Sue Doyle, *L.A.'s Trash Goal: No Waste by 2030*, DAILY BREEZE.COM, Jan. 21, 2008

Driving to Green Buildings: The Transportation Energy Intensity of Buildings, ENVTL. BUILDING NEWS, Sept. 2007, *available at* BuildingGreen.com, *at* http://www.buildinggreen.com/auth/article.cfm?filename=160901a.xml&printable=yes

MARK E. EBERHART, FEEDING THE FIRE: THE LOST HISTORY & UNCERTAIN FUTURE OF MANKIND'S ENERGY ADDICTION (2007)

Jae Edmonds et al., International Emissions Trading, *in* Climate Change: Science, Strategies, & Solutions 245 (Eileen Claussen, Vicki Arroyo Cochran & Debra P. Davis eds. 2001)

PAUL R. EHRLICH, THE POPULATION BOMB: POPULATION CONTROL OR RACE TO OBLIVION? 34 (1968)

Joel B. Eisen, *"Brownfields of Dreams"?: Challenges and Limits of Voluntary Cleanup Programs and Incentives*, 1996 U. ILL. L. REV. 883 (1996)

D.L. ELLIOT ET AL, AN ASSESSMENT OF THE AVAILABLE WINDY LAND AREA AND WIND ENERGY POTENTIAL IN THE CONTIGUOUS UNITED STATES (1991)

Chad D. Emerson, *Making Main Street Legal Again: The SmartCode Solution to Sprawl*, 71 Mo. L. REV. 637 (2006)

Emily, *Chicago Green Roof Program*, INHABITAT, Aug. 1, 2006, available at http://www.inhabitat.com/2006/08/01/chicago-green-roof-program

Kirsten H. Engel, *Harmonizing Regulatory and Litigation Approaches to Climate Change Mitigation: Incorporating Tradable Emissions Offsets Into Common Law Remedies*, 155 U. PA. L. REV. 1563 (2007)

Kirsten Engel, *State and Local Climate Change Initiatives: What is Motivating State and Local Governments to Address a Global Problem and What Does This Say About Federalism and Environmental Law?*, 38 URB. LAW. 1015 (2006)

Peter D. Enrich, *Business Tax Incentives: A Status Report*, 34 URB. LAW. 415 (2002)

EU CLIMATE CHANGE POLICY: THE CHALLENGE OF NEW REGULATORY INITIATIVES (Marjan Peeters & Kurt Deketelaere eds. 2006)

Ambrose Evans-Pritchard, *Why the Price of 'Peak Oil' is Famine*, TELEGRAPH.CO.UK, Feb. 9, 2008, *available at* http://www.telegraph.co.uk/money/main.jhtml?xml=/money/2008/02/07/cnoil107.xml

REID EWING ET AL., GROWING COOLER: THE EVIDENCE ON URBAN DEVELOPMENT AND CLIMATE CHANGE (2007)

EXECUTIVE SUMMARY, SURFACE TEMPERATURE RECONSTRUCTIONS FOR THE LAST 2,000 YEARS, National Academy of Sciences, June 22, 2006, *available at* http:www.nap.edu/catalogue/11676.html

Daniel A. Farber, *Adapting To Climate Change: Who Should Pay*, 23 J. LAND USE & ENVTL. L. 1 (2007)

Daniel A. Farber, *Basic Compensation for Victims of Climate Change*, 155 U. PA. L. REV. 1605 (2007)

Daniel A. Farber et al., *Reinventing Flood Control*, 81 TUL. L. REV. 1085 (2007)

Michael G. Faure, *Insurability of Damage Caused by Climate Change: A Commentary*, 155 U. PA. L. REV. 1875 (2007)

Fed. Emergency Mgmt. Agency, About FEMA: What We Do, http://www.fema.gov/about/what.shtm (last visited Apr. 1, 2007)

FEDERAL LAND MANAGEMENT AGENCIES (Pamela D. Baldwin ed. (2005)

Feds Allowing Tarsands to Become 'Most Destructive Project on Earth': Report, CBC NEWS, Feb. 15, 2008

Jennifer Felten, *Brownfield Redevelopment 1995–2005: An Environmental Justice Success Story?* 40 REAL PROP. PROB. & TR. J. 679 (2006)

Steven Ferrey, *Converting Brownfield Environmental Negatives into Energy Positives*, 34 B.C. ENVTL. AFF. L. REV. 417 (2007)

Steven Ferrey, *Nothing But Net: Renewable Energy and the Environment, Mid-American Legal Fictions, and Supremacy Doctrine*, 14 Duke Envtl. L. & Pol'y F. 1 (2003)

Steven Ferrey, *Sustainable Energy, Environmental Policy, and States' Rights: Discerning the Energy Future Through the Eye of the Dormant Commerce Clause*, 12 N.Y.U. ENVTL. L.J. 507 (2004)

C.B. FIELD ET AL., CONFRONTING CLIMATE CHANGE IN CALIFORNIA: ECOLOGICAL IMPACTS ON THE GOLDEN STATE (1999)

Fire Up the Microwave, AOL.COM. Sept. 2, 2007, at 2, *available at* http://coaches.aol.com/ kids-and-family/kostigen-rogers/go-green-save-money

Umbra Fisk, *The Environmentalist's New Clothes*, GRIST, July 12, 2004, *available at* http://www.grist.org/advice/ask/2004/07/12/umbra-clothing/index.html

TIM FLANNERY, THE WEATHERMAKERS: HOW MAN IS CHANGING THE CLIMATE AND WHAT IT MEANS FOR LIFE ON EARTH (2005)

FLEXIBILITY IN CLIMATE POLICY: MAKING THE KYOTO MECHANISMS WORK (Tim Jackson *et al.* eds. 2001)

Sheila R. Foster, *The City as an Ecological Space: Social Capital and Urban Land Use*, 82 NOTRE DAME L. REV. 527 (2006)

France Slaps Penalties on Gas-Guzzling Cars, YahooGreen, Dec. 5, 2007, *available at* http://green.yahoo.com/news/afp/20071205/lf_afp/franceclimatetransporttax.html

Kenneth D. Frederick & Peter H. Gleick, *Potential Impacts on U.S. Water Resources, in* Climate Change: Science, Strategies, & Solutions 63 (Eileen Claussen, Vicki Arroyo Cochran & Debra P. Davis eds. 2001)

THOMAS L. FRIEDMAN, THE WORLD IN FLAT: A BRIEF HISTORY OF THE TWENTY-FIRST CENTURY (2005)

ROBERT H. FREILICH, FROM SPRAWL TO SMART GROWTH: SUCCESSFUL LEGAL, PLANNING, AND ENVIRONMENTAL SYSTEMS (1999)

Roy Fuller, *Wind Energy Development on BLM Lands*, 24 J. LAND RESOURCES & ENVTL. L. 613 (2004)

George Galster et al, *Wrestling Sprawl to the Ground: Defining and Measuring an Elusive Concept*, 12 HOUS. POL'Y DEBATE 681 (2001)

Laura L. Garcia, *The United States Should Not Support Family Planning Services in Developing Nations, in* DEVELOPING NATIONS 124 (Berna Miller & James D. Torr eds. 2003)

E.L. Gaston, *Taking the Gloves Off of Homeland Security: Rethinking the Federal Framework for Responding to Domestic Emergencies*, 1 HARV. L. & POL'Y REV. 519 (2007)

ROSS GELBSPAN, BOILING POINT: HOW POLITICIANS, BIG OIL AND COAL, JOURNALISTS, AND ACTIVISTS ARE FUELING THE CLIMATE CRISIS—AND WHAT WE CAN DO TO AVERT DISASTER (2004)

ROSS GELBSPAN, THE HEAT IS ON: THE HIGH STAKES BATTLE OVER EARTH'S THREATENED CLIMATE (1997)

Amelia Gentleman, *Architects Aren't Ready for an Urbanized Planet*, INT'L HERALD TRIBUNE, Aug. 20, 2007), *available at* http://www.iht.com/bin/print.php?id+7182262

Jon Gertner, *The Future is Drying Up*, N.Y. TIMES, Oct. 21, 2007

Emily Gertz, *Tempting Fate: Fifteen Years After the Great Flood of 1993, Floodplain Development is Booming*, GRIST, Mar. 19, 2008, *available at* http://www.grist.org/feature/2008/03/19/gertz/

Breanne Gilpatrick, *Cities Push Tap Water as 'Better Than Bottled,'* MIAMI HERALD.COM, Oct. 11, 2007, *available at* http://www.miamiherald.com/news/miami_dade/v-print/story/267546.html

Timothy H. Gillis, *Sixth Circuit Bans Ohio Tax Credit Under the Commerce Clause, Casting a Pall on Incentives*, 101 J. TAX'N 359 (2004)

PAUL GIPE, WIND POWER: RENEWABLE ENERGY FOR HOME, FARM, AND BUSINESS (2004)

Robert L. Glicksman, *Global Climate Change and the Risks to Coastal Areas from Hurricanes and Rising Sea Levels: The Costs of Doing Nothing*, 52 LOY. L. REV. 1127 (2006)

GLOBAL DEVELOPMENT OF ORGANIC AGRICULTURE: CHALLENGES AND PROSPECTS (Niels Halberg et al. eds. 2006)

GLOBAL ENVIRONMENTAL OUTLOOK (GEO_4), *available at* http://www.unep.org/ geo/geo4/report/GEO-4_Report_Full_en.pdf

Global Warming, BUSINESS WEEK ON LINE, Aug. 16, 2004

LEIGH GLOVER, POSTMODERN CLIMATE CHANGE (2006)

Go Green Save Money, AOL.COM. Sept. 2, 2007, *available at* http://coaches.aol.com/kids-and-family/kostigen-rogers/go-green-save-money

Donald M. Goldberg & Angela Delfino, *The Impact of the Kyoto Protocol on U.S. Business, in* GLOBAL CLIMATE CHANGE AND U.S. LAW 101 (Michael B. Gerrard ed. 2007)

DAVID B. GOLDSTEIN, SAVING ENERGY GROWING JOBS: HOW ENVIRONMENTAL PROTECTION PROMOTES ECONOMIC GROWTH, PROFITABILITY, INNOVATION, AND COMPETITION (2007)

James K. Gooch, *Fenced In: Why* Sheff v. O'Neill *Can't Save Connecticut's Inner City Schools*, 22 QLR 395, 397 (2004)

Jeff Goodell, Big Coal: The Dirty Secret Behind America's Energy Future (2006)

Jeff Goodell, *The Prophet of Climate Change: James Lovelock*, Rolling Stone.com, Oct. 17, 2007, available at http://www.rollingstone.com/politics/story/16956300/the_prophet_of_climate_change_james_lovelock

Jeff Goodell, *The Ethanol Scam: One of America's Biggest Political Boondoggles*, Rolling Stone.com, July 24, 2007, available at http://www.rollingstone.com/politics/story/15635751/the_ethanol_scam_one_of_americas_biggest_political_boondoggles

David Goodstein, Out of Gas: The End of the Age of Oil (2004)

Rebekah Gordon, *'Green' Builders May Get Fast Track*, Palo Alto Daily News, July 25, 2007, available at http://www.paloaltodailynews.com/article/2007-7-19-0719-smc-green

Ruth Gordon, *Climate Change and the Poorest Nations: Further Reflections on Global Inequality*, 78 U. Colo. L. Rev. 1559 (2007)

Al Gore, An Inconvenient Truth: A Planetary Emergency of Global Warming and What we Can do About It (2006)

Lawrence H. Goulder & Brian M. Nadreau, *International Approaches to Reducing Greenhouse Gas Emissions*, in Climate Change Policy: A Survey 115 (Stephen H. Schneider, Armin Rosencranz & John O. Niles eds. 2002)

Sara Elizabeth Graditor, Comment, *Responsibility for the Restoration of the Hurricane Insurance Industry: Business Proposal or State Solution?* 31 Nova L. Rev. 527 (2007)

Michael Greenberger, *Preparing Vulnerable Populations for a Disaster: Inner-City Emergency Preparedness—Who Should Take the Lead?*, 10 DePaul J. Health Care L. 291 (2007), available at http://ssrn.com/abstract=1017887

Michael S. Greve & Jonathan Klick, *Preemption in the Rehnquist Court: A Preliminary Empirical Assessment*, 14 Sup. Ct. Econ. Rev. 43 (2006)

Michael Grunwald, *Cry Me a River*, Grist, Mar. 18, 2008, available at http://www.grist.org/feature/2008/03/18/grunwald/index.html

Michael Grunwald, *Setting the Stage for More Katrinas*, Time.com, Aug. 2, 2007, available at http://www.time.com/time/printout/0,8816,1649403,00.html

Subhrajit Guhathakurta & Patricia Gober, *The Impact of the Phoenix Urban Heat Island on Residential Water Use*, 73 J. Am. Planning Ass'n 317 (2007)

Peter Gumbel, *Pasta Panic: The Price of Wheat is Up 60% This Year, and in Italy They're Taking to the Streets Over the Cost of Tortellini*, 156 FORTUNE 47 (2007)

Gary S. Guzy, *Insurance and Climate Change*, in GLOBAL CLIMATE CHANGE AND U.S. LAW 541 (Michael B. Gerrard ed. 2007)

PETER HALL, GREAT PLANNING DISASTERS (1980)

GRAHAM HALLETT, THE SOCIAL ECONOMY OF WEST GERMANY (1973)

Charles J. Hanley, *'Drilling Up' Into Space for Energy*, ASSOCIATED PRESS, Dec. 23, 2007, available at http://ap.google.com/article/ALeqM5gMOg-D8-UyHFE3GmhgU 5eUMRVF0gD8TNBC2G0

James Hansen, *The Threat to the Planet*, NEW YORK REVIEW OF BOOKS at 53 (No. 12, July 13, 2006)

Kashif Haque, Note, *Internal Revenue Code Section 1989, the Tax Incentive for Brownfield Redevelopment: A Sheep in Wolf's Clothing*, 8 WASH. U. J. L. & POL'Y 371 (2002)

Scott W. Hardt, *Federal Land Management in the Twenty-First Century: From Wise Use to Wise Stewardship*, 18 HARV. ENVTL. L. REV. 345 (1994)

Kevin T. Haroff & Katherine Kirwan Moore, *Global Climate Change and the National Environmental Policy Act*, 42 U.S.F.L. REV. 155 (2007)

Christina K. Harper, *Climate Change and Tax Policy*, 30 B.C. INT'L & COMP. L. REV. 411 (2007)

Alexandra R. Harrington, *Presidential Powers Revisited: an Analysis of the Constitutional Powers of the Executive and Legislative Branches over the Reorganization and Conduct of the Executive Branch*, 44 WILLAMETTE L. REV. 63 (2007)

Paul G. Harris, *Collective Action on Climate Change: The Logic of Regime Failure*, 47 NAT. RESOURCES J. 195 (2007)

Wylie Harris, *Lawn to Farm: Suburbia's Silver Lining*, COMMON DREAMS.ORG, Jan. 24, 2008

DEAN HAWKES & WAYNE FORSTER, ENERGY EFFICIENT BUILDINGS: ARCHITECTURE, ENGINEERING, AND ENVIRONMENT (2002)

David J. Hayes & Joel C. Beauvais, *Carbon Sequestration*, in GLOBAL CLIMATE CHANGE AND U.S. LAW 691 (Michael B. Gerrard ed. 2007)

ARTHUR HEARNDEN, EDUCATION, CULTURE, AND POLITICS IN WEST GERMANY (1976)

RICHARD HEINBERG, POWER DOWN: OPTIONS AND ACTIONS FOR A POST-CARBON WORLD (2004)

B. Timothy Heinmiller, *The Politics of "Cap and Trade" Policies*, 47 NAT. RESOURCES J. 445 (2007)

Lisa Heinzerling, *Climate Change and the Clean Air Act*, 42 U.S.F.L. Rev. 111 (2007)

Lisa Heinzerling, *Climate Change, Human Health, and the Post-Cautionary Principle*, SSRN, Georgetown University, O'Neill Institute for National and Global Health Law Scholarship (Research Paper No. 4 Sept. 2007) (forthcoming Georgetown Law Journal), available at SSRN: http://ssrn.com/abstract=1008923 and BePress: http

Lisa Heinzerling & Frank Ackerman, *Law and Economics for a Warming World*, 1 Harv. L. & Pol'y Rev. 331 (2007)

Robert Henson, The Rough Guide to Climate Change: The Symptoms, The Science, The Solutions (2006)

Joni Hersch & W. Kip Viscusi, *Allocating Responsibility for the Failure of Global Warming Policies*, 155 U. Pa. L. Rev. 1657 (2007)

Jason Hill et al., *Environmental, Economic, and Energetic Costs and Benefits of Biodiesel and Ethanol Biofuels*, 103 Proc. Nat'l Acad. Sci. 11,206 (2006)

Mayer Hillman et al, The Suicidal Planet: How to Prevent Global Climate Catastrophe (2007)

Dilip Hiro, Blood of the Earth: The Battle for the World's Vanishing Oil Resources (2007)

Dennis Hirsch et al., *Emissions Trading—Practical Aspects*, in Global Climate Change and U.S. Law 627 (Michael B. Gerrard ed. 2007)

David Hodas, *State Initiatives*, in Global Climate Change and U.S. Law 343 (Michael B. Gerrard ed. 2007)

Denice Ferkick Hoffman & Barbara Coler, *Brownfields and the California Department of Toxic Substances Control: Key Programs and Challenges*, 31 Golden Gate U. L. Rev. 433 (2001)

James A. Holtkamp, *Dealing with Climate Change in the United States: The Non-Federal Response*, 27 J. Land Resources & Envtl. L. 79 (2007)

Lance Hosey, *Is Bigger Better?: New Research Supports Eco-Friendly Urbanism*, Architect Magazine, Aug. 1, 2007

In Hot Seat, Bush Unveils New Climate Strategy, msnbc.com, Jun. 1, 2007

John Houghton, Global Warming: The Complete Briefing (2d ed. 1997)

Wayne Hsiung & Cass R. Sunstein, *Climate Change and Animals* 155 U. Pa. L. Rev. 1695 (2007)

David Hughes & Martin Morgan Taylor, *And Can't Look Up and See the Stars*, 16 J. Envtl. L. 215 (2004)

David Hunter & James Salzman, *Negligence in the Air: The Duty of Care in Climate Change Litigation*, 155 U. Pa. L. Rev. 1740 (2007)

Ann Hwang, *The United States Should Support Family Planning Services in Developing Nations*, in DEVELOPING NATIONS 102 (Berna Miller & James D. Torr eds. 2003)

In Hot Seat, Bush Unveils New Climate Strategy, MSNBC.COM, Jun. 1, 2007

Response to Governor's Sustainable Energy Plan for the State of Illinois, INNOVATION IN ENERGY TECHNOLOGY: COMPARING NATIONAL INNOVATION SYSTEMS AT THE SECTORAL LEVEL (2006)

INTERGOVERNMENTAL PANEL ON CLIMATE CHANGE, CLIMATE CHANGE 2007, THE PHYSICAL SCIENCE BASIS, SUMMARY FOR POLICY MAKERS (2007), *available at* http://www.ipcc.ch/

INTERGOVERNMENTAL PANEL ON CLIMATE CHANGE, CLIMATE CHANGE 2001, *available at* http://www.grida.no/climate/ipcc_tar/wg1/index.htm

INTERNATIONAL CITY/COUNTY MANAGEMENT ASS'N, MEASURING SUCCESS IN BROWNFIELDS REDEVELOPMENT PROGRAMS (2002)

Henry C. Jackson, *Farmer Take Another Look at Wind Energy*, MIAMI HERALD, Sept. 24, 2007, *available at* http://www.miamiherald.com/business/AP/story/248358.html

PETER JACQUES, GLOBALIZATION AND THE WORLD OCEAN (2006)

Mark Jaffe, *Global Warming?*, DENVER POST, Dec. 26, 2006

Dale Jamieson, *Ethics, Public Policy and Global Warming* in MORALITY'S PROGRESS (2003)

BRUCE E. JOHANSEN, GLOBAL WARMING IN THE 21ST CENTURY (2006) (3 volumes)

Alex Johnson, *Shining a Light on Hazards of Fluorescent Bulbs: Energy-Efficient Coils Booming, But Disposal of Mercury Poses Problems*, MSNBC.COM Mar. 19, 2008, *available at* http://www.msnbc.msn.com/id/23694819/

Jason Scott Johnston, *Desperately Seeking Numbers: Global Warming, Species Loss, and the Use and Abuse of Quantification in Climate Change Policy Analysis*, 155 U. PA. L. REV. 1901 (2007)

Nick Johnstone, *Tradable Permits for Climate Change: Implications for Compliance, Monitoring, and Enforcement*, in CLIMATE-CHANGE POLICY 238 (Dieter Helm ed.2005)

Carolyn Jones, *It Won't Be Easy Being Green: Berkeley Sets Tough Course for its Residents to Follow to Help Reduce Emissions of Greenhouse Gases in City*, SAN FRANCISCO CHRONICLE, May 24, 2007, at A1

Patrik Jones, *It Won't Be Easy Being Green: Berkeley Sets Tough Course for its Residents to Follow to Help Reduce Emissions of Greenhouse Gases in City*, CHRISTIAN SCIENCE MONITOR, Nov. 27, 2007, *available at* http://www.csmonitor.com/2007/1127/p03s03-usgn.htm

Jonathan Jorissen, Note, *Katrina's House: The Constitutionality of the Forced Removal of Citizens from their Homes in the Wake of Natural Disasters*, 5 AVE MARIA L. REV. 587 (2007)

MATTHEW E. KAHN, GREEN CITIES: URBAN GROWTH AND THE ENVIRONMENT (2006)

Jocelyn Kaiser, *Money—With Strings—to Fight Poverty*, 319 SCIENCE 754 (Feb. 2008)

Elaine C. Kamarck, *When First Responders Are Victims: Rethinking Emergency Response*, 1 HARV. L. & POL'Y REV. 97 (2007)

Anya Kamenetz, *The Green Standard?: LEED Buiuldings Get Lots of Buzz, But the Point is Getting Lost*, FAST COMPANY, *available at* http//www.fastcompany.com/magazine/119/the-green-standard_Printer_Friendly.html

Alice Kaswan, *The Domestic Response to Global Climate Change: What Role for Federal, State, and Litigation Initiatives?*, 42 U.S.F.L. REV. 39 (2007)

Sajadi, Keeana, Comment, *The Terminator a Trendsetter? How California's Global Warming Solutions Act Will Impact California, the United States, and the World*, 21 J. NAT. RESOURCES & ENVTL. L. 143 (2006–2007)

DOUGLAS KELBAUGH, COMMON PLACE: TOWARD NEIGHBORHOOD AND REGIONAL DESIGN (1997)

Richard A. Kerr *et al.*, *Latest Forecast: Stand By for a Warmer But not Scourching, World*, 312 SCIENCE 351, 351 (2006)

Charles J. Kibert & Kevin Grosskopf, *Envisioning Next-Generation Green Buildings*, 23 J. LAND USE & ENVTL. L. 145 (2007)

Nancy J. King & Brian J. King, *Creating Incentives for Sustainable Buildings: A Comparative Law Approach Featuring the United States and the European Union*, 23 VA. ENVTL. L.J. 397 (2005)

Benjamin S. Kingsley, *Making it Easy to be Green: Using Impact Fees to Encourage Green Building*, 83 N.Y.U. L. REV. 1 (2008) (draft version)

Christine A. Klein, *The New Nuisance: An Antidote to Wetland Loss, Sprawl, and Global Warming*, 48 B.C.L. Rev. 1155 (2007)

Jeffrey Kluger, *Global Warming Heats Up*, TIME, Mar. 26, 2006)

Gerrit-Jan Knaap, & John W. Frece, *Smart Growth in Maryland: Looking Forward and Looking Back*, 43 IDAHO L. REV. 445 (2007)

J.B. KNOX & A. FOLEY, GLOBAL CLIMATE CHANGE AND CALIFORNIA: POTENTIAL IMPACTS AND RESPONSES (1992)

Daisuke Kojo, *The Importance of the Geographic Origin of Agricultural Products: A Comparison of Japanese and American Approaches*, 14 MO. ENVTL. L. & POL'Y REV. 275 (2006)

Laura H. Kosloff & Mark C. Trexler, *Consideration of Climate Change in Facility Permitting*, in GLOBAL CLIMATE CHANGE AND U.S. LAW 259 (Michael B. Gerrard ed. 2007)

TONY KOSLOW, THE SILENT DEEP: THE DISCOVERY, ECOLOGY AND CONSERVATION OF THE DEEP SEA (2007)

Howard C. Kunreuther & Erwann O. Michel-Kerjan, *Climate Change, Insurability of Large-Scale Disasters, and the Emerging Liability Challenge*, 155 U. PA. L. REV. 1795 (2007)

JAMES HOWARD KUNSTLER, THE LONG EMERGENCY: SURVIVING THE CONVERGING CATASTROPHES OF THE TWENTY-FIRST CENTURY (2005)

Julianne Kurdila & Elise Rindfleisch, *Funding Opportunities for Brownfield Development*, 34 B.C. ENVTL. AFF. L. REV. 479 (2007)

James A. Kushner, *Brownfield Redevelopment Strategies in the United States*, 22 GA. ST. U. L. REV. 857 (2006)

James A. Kushner, *Car Free Housing Developments: Towards Sustainable Smart Growth and Urban Regeneration Through Car-Free Zoning, Car-Free Redevelopment, Pedestrian Improvement Districts, and New Urbanism*, 23 UCLA J. Envtl. L. & Pol'y 1 (2005)

JAMES A. KUSHNER, THE POST-AUTOMOBILE CITY (2004)

James A. Kushner, *Growth for the Twenty-First Century: Tales from Bavaria and the Vienna Woods—Comparative Images of Urban Planning in Munich, Salzburg, Vienna, and the United States*, 29 URB. LAW. 911, 924 (1997), *reprinted as modified*, 6 S. CAL. INTERDISC. L.J. 89 (1997)

JAMES A. KUSHNER, HEALTHY CITIES—THE INTERSECTION OF URBAN PLANNING, LAW, AND HEALTH (2007)

JAMES A. KUSHNER, THE POST-AUTOMOBILE CITY (2004)

James A. Kushner, *Smart Growth, New Urbanism, and Diversity: Progressive Planning Movements in America and Their Impact on Poor and Minority Ethnic Populations*, 21 UCLA J. ENVTL L. & POL'Y 45 (2002/2003)

James A. Kushner, *Social Sustainability: Planning for Growth in Distressed Places—the German Experience in Berlin, Wittenberg, and the Ruhr*, 3 WASH. U. J. L. & POL'Y 849 (2000), *published in* EVOLVING VOICES IN LAND USE LAW ch. 13 (Wash. U. J. L. & Pol'y ed., 2000)

1 JAMES A. KUSHNER, SUBDIVISION LAW AND GROWTH MANAGEMENT (2d ed. 2001 & Supp. 2007)

Richard D. Lamm, *Immigration: The Ultimate Environmental Issue*, 84 DENV. U. L. REV. 1003 (2007)

CAROL LANCASTER, TRANSFORMING FOREIGN AID: UNITED STATES ASSISTANCE IN THE 21ST CENTURY (2000)

Robert E. Lang & Arthur C. Nelson, *America 2040: The Rise of the Megapolitans*, PLANNING, Jan. 2007

THE LAW OF THE SEA: PROGRESS AND PROSPECTS (David Freestone et al. eds 2006)

Lyndsey Layton & Spencer S. Hsu, *Letting the Market Drive Transportation: Bush Officials Criticized for Privatization*, WASHINGTONPOST.COM, Mar. 17, 2008, *available at* http://www.washingtonpost.com/wp-dyn/content/article/2008/03/16/AR2008031603085.html

Edith M. Lederer, *Biofuel Growth Adds to Hunger: Most Vulnerable 'Priced Out' of Market for Food*, Wash. Times, Feb. 14, 2008

Edith M. Lederer, *UN Expert Decries Turning Food Into Fuel*, YAHOO NEWS, Oct. 27, 2007, *available at* http://news.yahoo.com/s/ap/20071026/ap_on_re_us/un_food_vs_biofuel_3

A LEGAL GUIDE TO HOMELAND SECURITY AND EMERGENCY MANAGEMENT FOR STATE AND LOCAL GOVERNMENTS (Ernest B. Abbott & Otto J. Hetzel eds. 2005)

Andrew Leonard, *A Plague of Bloodsuckers: In Japan, Reforestation, Population Decline and Global Warming Have Set Off a Land Leech Invasion*, SALON.COM, Sept. 7, 2007, *available at* http://www.salon.com/tech/htww/2007/09/07/a_plague_leeches/print.html

MARCEL LEROUX, GLOBAL WARMING — MYTH OR REALITY? THE ERRING WAYS OF CLIMATOLOGY (2005)

Michael H. LeRoy, *Compulsory Labor in a National Emergency: Public Service or Involuntary Servitude? The Case of Crippled Ports*, 28 BERKELEY J. EMP. & LAB. L. 331 (2007)

Michael Lewyn, *How Government Regulation Forces Americans into their Cars: A Case Study*, 16 WIDENER L.J. 839 (2007)

Michael Lewyn, *Sprawl, Growth Boundaries and the Rehnquist Court*, 2002 UTAH L. REV. 1

Michael Lewyn & Shane Cralle, *Planners Gone Wild: The Overregulation of Parking*, 33 WM. MITCHELL L. REV. 613 (2007)

Lib Dems Plan Air Tax to aid rail, BBC News, Aug. 2, 2007, available at http://newsvote.bbc.co.uk/mpapps/pagetools/print/news.bbc.co.uk/2/hi/uk_news/politics/69

Allison Linn, *Carbon Offset Market Raises Questions*, MSNBC, May 22, 2007, available at http://www.msnbc.msn.com/id/18659716/

Goodwin Liu, *Education, Equality, and National Citizenship*, 116 YALE L.J. 330 (2006)

MASSIMO LIVI-BACCI: A CONCISE HISTORY OF WORLD POPULATION (4th ed. 2007)

ALAN LONGHURST, ECOLOGICAL GEOGRAPHY OF THE SEA (2d ed.2007)

Orie L. Loucks, *Business Capitalizing on Energy Transition Opportunities*, in CLIMATE CHANGE POLICY: A SURVEY 495 (Stephen H. Schneider, Armin Rosencranz & John O. Niles eds. 2002)

JAMES LOVELOCK, GAIA: A NEW LOOK AT LIFE ON EARTH (2000)

JAMES LOVELOCK, HEALING GAIA: PRACTICAL MEDICINE FOR THE PLANET (1991)

Amory B. Lovins, *Winning the Oil Endgame: Innovation for Profits, Jobs, and Security*, 84 FOREIGN AFFAIRS 152 (2004)

Marcia D. Lowe, *Cars, Their Problems, and the Future*, in CITIES AND CARS 221 (Roger L. Kemp ed. 2007)

Karen MacDonald & Zen Makuch, *Emissions Trading and the Aarhus Convention: A Proportionate Symbiosis?*, in EU CLIMATE CHANGE POLICY: THE CHALLENGE OF NEW REGULATORY INITIATIVES 125 (Marjan Peeters & Kurt Deketelaere eds. 2006)

William P. Macht, *The Rise of Car Sharing*, 62 URB. LAND 26 (Jan. 2003)

Andrew Manale, *Agriculture and the Developing World: Intensive Animal Production, a Growing Environmental Problem*, 19 GEO. INT'L ENVTL. L. REV. 809 (2007)

David G. Mandelbaum, *Corporate Sustainability Strategies*, 26 TEMP. J. SCI. TECH. & ENVTL. L. 27 (2007)

Jerry Mander, *Globalization is Harmful to the Environment*, in GLOBALIZATION: OPPOSING VIEWPOINTS 84 (Louise I. Gerdes ed. 2006)

Manhattan Parking Spot Going for $225,000, CNNMONEY.COM, July 12, 2007

Bradford C. Mank, *Reforming State Brownfield Programs to Comply with Title VI*, 24 HARV. ENVTL. L. REV. 115 (2000)

Roberta Mann, *Another Day Older and Deeper in Debt: How Tax Incentives Encourage Burning Coal and the Consequences for Global Warming*, 20 GLOBAL BUS. & DEV. L. J. 111 (2007)

Roberta Mann, *Subsidies, Tax Policy, and Technological Innovation*, in GLOBAL CLIMATE CHANGE AND U.S. LAW 565 (Michael B. Gerrard ed. 2007)

RICHARD MANNING, AGAINST THE GRAIN: HOW AGRICULTURE HAS HIJACKED CIVILIZATION (2004)

Anne Marie Mannion, *In it for the Short Haul: Car-Sharing for Urban Errands Brings it to Battle Against Price, Parking and Pollution to Chicago*, CHI. TRIB., Sept. 12, 2002 at N1

Robert M. Margolis & Daniel M. Kammen, *Energy R&D and Innovation: Challenges and Opportunities*, in CLIMATE CHANGE POLICY: A SURVEY 469 (Stephen H. Schneider, Armin Rosencranz & John O. Niles eds. 2002)

Simon Marr, Precautionary Principle in the Law of the Sea: Modern Decisionmaking in International Law (2003)
James Martin, The Meaning of the 21st Century (2006)
Mark Martin, *Sprawl Clashes with Warming in California*, San Francisco Chronicle, May 27, 2007
Aileen M. Marty, *Hurricane Katrina: A Deadly Warning Mandating Improvement to the National Response to Disasters*, 31 Nova L. Rev. 423 (2007)
Paul Mason, Planet Under Pressure—Population (2006)
Mark Mazzetti & Joel Havemann, *Iraq War is Costing $100,000 per Minute*, L.A. Times, Feb. 3, 2006
Thomas H. Maugh II & Karen Kaplan, *Katrina Leaves Permanent Scar on Forests*, L.A. Times, Nov. 16, 2007, *available at* http://www.latimes.com/news/science/la-sci-trees16nov16,1,2189243.story
J. Michael McConnell, *Annual Threat Assessment of the Director of National Intelligence*, Feb. 5, 2008, *available at* http://www.tsa.gov/assets/pdf/02052008_dni_testimony.pdf
Karen MacDonald & Zen Makuch, *Emissions Trading and the Aarhus Convention: A Proportionate Symbiosis?*, *in* EU Climate Change Policy: The Challenge of New Regulatory Initiatives 125 (Marjan Peeters & Kurt Deketelaere eds. 2006)
George C. McGavin, Endangered: Wildlife on the Brink of Extinction (2006)
Bill McKibben, Hope, Human and Wild: True Stories of Living Lightly on the Earth (1995)
Robin McKie, *How Africa's Desert Sun Can Bring Europe Power*, The Observer, Dec. 2, 2007, *available at* http://www.guardian.co.uk/environment/2007/dec/02/renewableenergy.solarpower/
Robert B. McKinstry, Jr., *Laboratories for Local Solutions for Global Problems: State, Local and Private Leadership in Developing Strategies to Mitigate the Causes and Effects of Climate Change*, 12 Penn St. Envtl. L. Rev. 15 (2004)
Amy Pilat McMorrow, Note, *CERCLA Liability Redefined: An Analysis of the Small Business Liability Relief and Brownfields Revitalization Act and its Impact on State Voluntary Cleanup Programs*, 20 Ga. St. U. L. Rev. 1087 (2004)
Sheila McNulty, *Green Leaves, Black Gold*, Financial Times, Dec. 16, 2007
Frederick A. B. Meyerson, *Population and Climate Change Policy*, *in* Climate Change Policy: A Survey 251 (Stephen H. Schneider, Armin Rosencranz & John O. Niles eds. 2002)

Alan S. Miller, *International Trade and Development*, in GLOBAL CLIMATE CHANGE AND U.S. LAW 277 (Michael B. Gerrard ed. 2007)

Kathleen A. Miller, *Climate Change and Water in the West: Complexities, Uncertainties and Strategies for Adaptation*, 27 J. LAND RESOURCES & ENVTL. L. 87 (2007)

James F. Miskel, Disaster Response and Homeland Security: What Works, What Doesn't (2006)

Bruce Mohl, *More Insurers Backing Away From Coasts*, BOSTON GLOBE, Oct. 14, 2007, available at http://www.boston.com/business/personalfinance/articles/2007/10/14/more_insurers_bac

GEORGE MONBIOT, HEAT: HOW TO STOP THE PLANET FROM BURNING (2007)

THEODORE H. MORAN, HARNESSING FOREIGN DIRECT INVESTMENT FOR DEVELOPMENT: POLICIES FOR DEVELOPED AND DEVELOPING COUNTRIES (2006)

Jennifer P. Morgan, Note, *Carbon Trading Under the Kyoto Protocol: Risks and Opportunities for Investors*, 18 FORDHAM ENVTL. L. REV. 151 (2006)

CRAIG MORRIS, ENERGY SWITCH: PROVEN SOLUTIONS FOR A RENEWABLE FUTURE (2006)

Julian Morris, *Warming Aid, Chilling Trade?*, in ADAPT OR DIE: THE SCIENCE, POLITICS AND ECONOMICS OF CLIMATE CHANGE 133 (Kendra Okonski ed. 2003)

Siwa Msangi & Mark Rosegrant, *Agriculture and the Environment: Linkages, Trade-offs and Opportunities*, 19 GEO. INT'L ENVTL. L. REV. 699 (2007)

Msnbc, *U.N. Issues Landmark Report on Global Warming: Panel Offers Dire Warnings, Establishes Scientific Baseline for Political Talks*, MSNBC, Nov. 17, 2007, available at http://www.msnbc.msn.com/id/21844627/ (Synthesis Report)

THE MULTI-GOVERNANCE OF WATER: FOUR CASE STUDIES (Matthias Finger et al., eds. 2005)

Haya El Nasser, *Senior Transportation a Growing Concern*, USA TODAY, Dec. 2, 2007, available at http://www.usatoday.com/news/nation/2007-12-02-transport_N.htm

National Ass'n of Realtors, News Release, *Americans Prefer to Spend More on Mass Transit and Highway Maintenance, Less on New Roads*, Oct. 24, 2007, available at http://www.realtor.org/press_room/news_releases/2007/nar_smart_growth_survey_2007.html

Laura Navarro, Comment, *What About the Polar Bears? The Future of the Polar Bears as Predicted by a Survey of Success under the Endangered Species Act*, 19 VILL. ENVTL. L.J. 169 (2008)

Michael Neibauer, *Legislation to Ease Parking Crunch—for Bicycles*, D.C. Examiner, Sept. 13, 2007

Arthur C. Nelson, Casey J. Dawkins & Thomas W. Sanchez, The Social Impacts of Urban Containment (2007)

William Neuman, *Mixed Signals: Driving to Work as a Tax Break*, N.Y. Times, Aug. 16, 2007

New Jersey Town Doubles Recyling Rates in One Week with the RecycleBank Program, Green Progress, Dec. 3, 2007, *available at* http://www.enn.com/pollution/article/26368

New Virus Extends Geographic Reach, L.A. Times, Dec. 8, 2007, at A13

John O. Niles, *Tropical Forests and Climate Change*, *in* Climate Change Policy: A Survey ch. 13 (Stephen H. Schneider, Armin Rosencranz & John O. Niles eds. 2002)

John R. Nolan, *Disaster Mitigation Through Land Use Strategies*, 37 Envt'l L. Rptr. 10681 (2007)

Peter Carl Nordberg, Note, *Excuse Me, Sir, But Your Climate's on Fire: California's S.B. 1368 and the Dormant Commerce Clause*, 82 Notre Dame L. Rev. 2067 (2007)

John Norquist, *We Would Use Less Energy Living Closer Together: Cities Have Powerful Environmental Advantages: They Make it Easier to Walk and Use Public Transit*, philly.com, May 17, 2007, *available at* http:www.philly.com/inquirer/opinion/20070517_We_would_use_less_energy_living_closer_together.html

Note, *The Compact Clause and the Regional Greenhouse Gas Initiative*, 120 Harv. L. Rev. 1958 (2007)

Note, *Making Mixed-Income Communities Possible: Tax Base and Class Desegregation*, 114 Harv. L. Rev. 1575 (2001)

Stephanie B. Ohshita, *The Scientific and International Context for Climate Change Initiatives*, 42 U.S.F.L. Rev. 1 (2007)

Rachel Oliver, *All About: Cities and Energy Consumption*, CNN.com, Dec. 31, 2007, *available at* http://www.cnn.com/2007/TECH/12/31/eco.cities/index.html

Dave Olson, *Fare-Free Public Transit Could be Headed to a City Near You*, AlterNet, Aug. 2, 2007, *available at* http://www.alternet.org/module/printversion/57802

Charles Openchowski, *The Next Greenhouse Gas Executive Order?*, 38 Envtl. L. Rep. News & Analysis 10077 (2008)

Richard G. Opper, *The Brownfield Manifesto*, 37 Urb. Law. 163 (2005)

Naomi Oreskes, *Beyond the Ivory Tower: The Scientific Consensus on Climate Change*, 306 Science 1686, 1686 (2004)

Naomi Oreskes, *The Scientific Consensus on Climate Change: How Do We Know We're Not Wrong?*, in CLIMATE CHANGE: WHAT IT MEANS FOR US, OUR CHILDREN, AND OUR GRANDCHILDREN 65 (Joseph F.C. DiMento & Pamela Doughman eds. 2007)

MYRON ORFIELD, AMERICAN METROPOLIS: THE NEW SUBURBAN REALITY (2002)

Hari M. Osofsky, *Local Approaches to Transnational Corporate Responsibility: Mapping the Role of Subnational Climate Change Litigation*, 20 GLOBAL BUS. & DEV. L. J. 143 (2007)

Pace Law School Center for Environmental Legal Studies, *State Response to Climate Change: 50-State Survey*, in GLOBAL CLIMATE CHANGE AND U.S. LAW 371 (Michael B. Gerrard ed. 2007)

GREG PAHL, THE CITIZEN-POWERED ENERGY HANDBOOK: COMMUNITY SOLUTIONS TO A GLOBAL CRISIS (2007)

Marc Pallemaerts & Rhiannon Williams, *Climate Change: The International and European Policy Framework,* in EU CLIMATE CHANGE POLICY: THE CHALLENGE OF NEW REGULATORY INITIATIVES 22, 37–41 (Marjan Peeters & Kurt Deketelaere eds. 2006)

Ian Parry & Elena Safirova, *Pay as You Slow: Road Pricing to Reduce Traffic Congestion,* in NEW APPROACHES ON ENERGY AND THE ENVIRONMENT: POLICY ADVICE FOR THE PRESIDENT 63 (Richard D. Morgenstern & Paul R. Portney eds. 2004)

Ian W.H. Parry, *Fiscal Interactions and the Case for Carbon Taxes Over Grandfathered Carbon Permits,* in CLIMATE-CHANGE POLICY 218 (Dieter Helm ed. 2005)

Holly L. Pearson, *Climate Change and Agriculture: Mitigation Options and Potential,* in CLIMATE CHANGE POLICY: A SURVEY 307 (Stephen H. Schneider, Armin Rosencranz & John O. Niles eds. 2002)

Marjan Peeters, *Enforcement of the EU Greenhouse Gas Emissions Trading Scheme,* in EU CLIMATE CHANGE POLICY: THE CHALLENGE OF NEW REGULATORY INITIATIVES 169–86 (Marjan Peeters & Kurt Deketelaere eds. 2006)

Neil Peirce, *Follow British Model on Transportation Needs: It's Time for a Big New Tax in America*, DAILYPRESS.COM, Aug. 19, 2007, *available at* dailypress.com/news/opinion/dp-op_peirce_0819aug19,0,1414318.story

Neal Peirce, *Study Shows High Sea Rise Danger for U.S. Coastal Cities*, POSTWRITERSGROUP.COM, Sept. 9, 2007, *available at* http://www.postwritersgroup.com/archives/peir070909.htm

Michael J. Percy, *Delta Levees—Tort Immunity vs. Takings Liability*, 42 REAL PROP. PROB. & TR. J. 547 (2007)

Mary E. Peters, *Gas Taxes Are High Enough*, WALL ST. J., Jan. 18, 2008, at A13, *available at* http://online.wsj.com/public/article_print/SB120062474267 899727.html.

PEW CENTER ON GLOBAL CLIMATE CHANGE, LEARNING FROM STATE ACTION ON CLIMATE CHANGE (Mar. 2006), *update available at* http://pewclimate.org/docUploads/UpdatePewStatesBriefMarch2006%2Epdf

Neal Peirce, *Study Shows High Sea Rise Danger for U.S. Coastal Cities*, POSTWRITERSGROUP.COM, Sept. 9, 2007, *available at* http://www.postwritersgroup.com/archives/peir070909.htm

John K. Pierre & Gail S. Stephenson, *After Katrina: A Critical Look at FEMA's Failure to Provide Housing for Victims of Natural Disasters*, 68 LA. L. REV. 443 (2008)

A. BARRIE PITTOCK, CLIMATE CHANGE: TURNING UP THE HEAT (2005)

RUTHERFORD PLATT, LAND USE AND SOCIETY: GEOGRAPHY, LAW AND PUBLIC POLICY (1996)

Kristen M. Ploetz, Note, *Light Pollution in the United States: An Overview of the Inadequacies of the Common Law and State and Local Regulation*, 36 NEW ENG. L. REV. 985 (2002)

THE POISONED WELL: NEW STRATEGIES FOR GROUNDWATER PROTECTION (Eric P. Jorgensen ed. 1989)

Gary A. Poliakoff, *Disaster Planning and Recovery*, 31 NOVA L. REV. 457 (2007)

Eric A. Posner, *Climate Change and International Human Rights Litigation: A Critical Appraisal*, 155 U. PA. L. REV. 1925 (2007)

SANDRA POSTEL, LAST OASIS: FACING WATER SCARCITY (1992)

Robert Preer, *Housing Deal Gets Popular: Towns Just Hope State Ponies Up Aid*, BOSTON GLOBE, Sept. 2, 2007, *available at* http://www.boston.com/news/local/articles/2007/09/02/housing_deal_gets_popular?mode=PF

SAMUEL T. PRESCOTT, FEDERAL LAND MANAGEMENT: CURRENT ISSUES AND BACKGROUND (2003)

Bill Prindle, *How Energy Efficiency Can Turn 1300 New Power Plants Into 170, fact sheet* (Wash. D.C. Alliance to Save Energy, May 2, 2001)

George (Rock) Pring, *A Decade of Emissions Trading in the USA: Experiences and Observations for the EU*, *in* EU CLIMATE CHANGE POLICY: THE CHALLENGE OF NEW REGULATORY INITIATIVES 188–201 (Marjan Peeters & Kurt Deketelaere eds. 2006)

PUBLIC LANDS: USE AND MISUSE (William E. Neeley ed. 2007)

Philip Pullella, *Save the Planet Before it's Too Late, Pope Urges*, YAHOO NEWS, Sept. 2, 2007, *available at* http://news.yahoo.com/s/nm/20070902/wl_nm/pope_dc_1;_ylt=Aq3_KwhkpAXqiTBnDtyJ67BkM3wV

GEORGE PYLE, RAISING LESS CORN, MORE HELL: THE CASE FOR THE INDEPENDENT FARM AND AGAINST INDUSTRIAL FOOD (2005)

Quebec to Collect Nation's 1st Carbon Tax: Energy Companies will Pass Cost to Consumers, Say Analysts, CBC NEWS, Jun. 7, 2007)

BARRY G. RABE, PEW CTR. ON GLOBAL CLIMATE CHANGE, RACE TO THE TOP: THE EXPANDING ROLE OF U.S. STATE RENEWABLE PORTFOLIO STANDARDS 3–4 (2006), *available at* http://www.pewclimate.org/docUploads/RPSReportFinal%2Epdf

BARRY G. RABE, STATEHOUSE AND GREENHOUSE: THE EMERGING POLITICS OF AMERICAN CLIMATE CHANGE POLICY (2004)

ROBERT RADVANOVSKY, CRITICAL INFRASTRUCTURE: HOMELAND SECURITY AND EMERGENCY PREPAREDNESS (2006)

Real Climate, http//www.realclimate.org

IRWIN REDLENER, AMERICANS AT RISK: WHY WE ARE NOT PREPARED FOR MEGADISASTERS AND WHAT WE CAN DO NOW (2006)

RICHARD REGISTER, ECOCITIES: REBUILDING CITIES IN BALANCE WITH NATURE (Rev. ed. 2006)

MARK REISCH & DAVID M. BEARDEN, SUPERFUND AND THE BROWNFIELDS ISSUE (2003)

MARC REISNER, CADILLAC DESERT: THE AMERICAN WEST AND ITS DISAPPEARING WATER (1993)

Lisa Rein, *Md. House Approves Cut in Car Pollution*, WASH. POST, Feb. 21, 2007, at A01

RENEWABLE ENERGY: POWER FOR A SUSTAINABLE FUTURE (2d ed. Godfrey Boyle ed. 2004)

RENEWABLE RESOURCES AND RENEWABLE ENERGY: A GLOBAL CHALLENGE (Mauro Graziani & Paolo Fornasiero eds. 2007)

Rethinking *the Kyoto Protocol: Are There Legal Solutions to Global Warming and Climate Change?*, 5 WASH. U. GLOBAL STUD. L. REV. 333 (2006)

Reverse Osmosis and Toray Membrane: Can Desalination Finally Solve Water Scarcity?, VERDEXCHANGE, Nov. 12, 2007, *available at* http://verdexchange.org/200710/reverse.html

Andrew C. Revkin, *Budgets Falling in Race to Fight Global Warming*, N.Y. TIMES, Oct. 30, 2006, at A1

Andrew C. Revkin, *Climate Change as News: Challenges in Communicating Environmental Science*, in CLIMATE CHANGE: WHAT IT MEANS FOR US, OUR CHILDREN, AND OUR GRANDCHILDREN 139 (Joseph F.C. DiMento

& Pamela Doughman eds. 2007)

J.B. Ruhl, *Climate Change and the Endangered Species Act: Building Bridges to the No-Analog Future*, 88 B.U. L. Rev. 1 (2008)

Gary Richards, *VTA Finds Hydrogen Buses Cost Much More to Run Than Diesel Vehicles*, MercuryNews.com, Feb. 26, 2008, available at http://www.mercurynews.com/ci_8365544

Jonathan Riker, *The Green Zone: Green Building Requirements Must Strike a Balance Between Market Economics and Social Needs*, 30 Los Angeles Law. 27 (Jan. 2008)

J.F. Rischard, High Noon: Twenty Global Problems, Twenty Years to Solve Them (2002)

Rising Sea Levels Send Ripples Through Real Estate Industry, San Diego Union Tribune, Jun. 24, 2007, available at http://www.signonsandiego.com/news/features/20070624-9999-lz1c24smokes.html

John Ritter, *Calif. Sees Sprawl as Warming Culprit*, USA Today, Jun. 14, 2007

Mildred Wigfall Robinson, *Fulfilling Brown's Legacy: Bearing the Costs of Realizing Equality*, 44 Washburn L.J. 1 (2004)

Elizabeth Rogers & Thomas Kostigen, The Green Book: The Everyday Guide to Saving the Planet One Simple Step at a Time (2007)

Joseph J. Romm, Hell and High Water: Global Warming—The Solution and the Politics—And What We Should Do (2007)

Joseph J. Romm, The Hype About Hydrogen: Fact and Fiction in the Race to Save the Climate (2005)

Michael S. Rosenwald, *The Rising Tide of Corn*, Wash. Post, Jun. 15, 2007, at D01

David Rusk, Inside Game—Outside Game: Winning Strategies for Saving Urban America (1999)

Russian Region to Host Day of Conception, Contra Costa Times, Sept. 11, 2007

Elisabeth Rosenthal, *U.N. Report Describes Risks of Inaction on Climate Change*, N.Y. Times, Nov. 17, 2007, available at http://www.nytimes.com/2007/11/17/science/earth/17climate.html?_r=1&ref=todayspaper&oref=slogin

James S. Russell, *Can LEED Survive the Carbon-Neutral Era?*, Metropolis Magazine, Nov. 2007, available at http://www.metropolismag.com/cda/print_friendly.php?artid=3051

James E. Ryan, *Schools, Race, and Money*, 109 Yale L.J. 249 (1999)

James E. Ryan & Michael Heise, *The Political Economy of School Choice*, 111 Yale L.J. 2043 (2002)

Sale of Carbon Credits Helping Land-Rich, But Cash-Poor, Tribes, N.Y. Times.com, May 7, 2007

Patricia E. Salkin, *Sustainability at the Edge: The Opportunity and Responsibility of Local Governments to Most Effectively Plan for Natural Disaster Mitigation*, 38 ENV'T L. RPTR. 10158 (Mar. 2008)
Peter W. Salsich, Jr., *Toward a Policy of Heterogeneity: Overcoming a Long History of Socioeconomic Segregation in Housing*, 42 WAKE FOREST L. REV. 459 (2007)
Ian Sample, *Global Food Crisis Looms as Climate Change and Population Growth Strip Fertile Land*, GUARDIAN, Aug. 31, 2007, available at http://www.guardian.co.uk/environment/2007/aug/31/climatechange.food.
SB 375 Connects Land Use and AB 32 Implementation, PLANNING REPORT, July, 2007), available at http://www.planningreport.com/tpr/?module=displaystory_id=1257&format=html
CHRISTIAN SCHABBEL, THE VALUE CHAIN OF FOREIGN AID: DEVELOPMENT, POVERTY REDUCTION, AND REGIONAL CONDITIONS (2007)
T.C. Schelling, *Economic Responses to Global Warming: Prospects for Cooperative Approaches*, in GLOBAL WARMING: ECONOMIC POLICY RESPONSES 197 (Rudiger Dornbusch & James M. Poterba eds. 1991)
Stephen H. Schelling, *Economic Responses to Global Warming: Prospects for Cooperative Approaches*, in CLIMATE CHANGE POLICY: A SURVEY 53 (Stephen H. Schneider, Armin Rosencranz & John O. Niles eds. 2002)
Stephen H. Schneider & Kristin Kuntz-Duriseti, *Uncertainty and Climate Change Policy*, in CLIMATE CHANGE POLICY: A SURVEY 53, 78 (Stephen H. Schneider, Armin Rosencranz & John O. Niles eds. 2002)
Michael H. Schuitema, Comment, *Road Pricing as a Solution to the Harms of Traffic Congestion*, 34 TRANSP. L.J. 81 (2007)
Anna K. Schwab & David Brower, *Increasing Resilience to Natural Hazards: Obstacles and Opportunities for Local Governments Under the Disaster Mitigation Act of 2000*, 38 ENV'T L. RPTR. 10171 (Mar. 2008)
Ariel R. Schwartz, Note, *Doubtful Duty: Physicians' Legal Obligation to Treat During an Epidemic*, 60 STAN. L. REV. 657 (2007)
Secretary Peters Says Bikes "Are Not Transportation," Excerpt from Interview on PBS "NewsHour" with Jim Lehrer, available at http:www.streetsblog.org/2007/08,17/secretary-Peters-says-bikes-are-not-transportation/
Select Bipartisan Comm. to Investigate the Preparation for and Response to Hurricane Katrina, A Failure of Initiative: Final Report, H.R. Rep. No. 109–377 (2006)
Robert Selna, *Eco-Tough S.F. Code Proposed for Buildings*, SFGATE.COM, July 11, 2007
James Shepherd, *The Future of Technology Transfer Under Multilateral Agreements*, 37 ENVTL L. RPTR. NEWS & ANALYSIS 10547 (2007)

Kate Sheppard, *The Urban Revival: Cities May be the Key to Curbing Climate Crisis*, MSN, Aug. 30, 2007, available at http://stopglobalwarming.msn.com/article.aspx?cp-documentid=5288633

Kristina Shevory, *Homespun Electricity, From the Wind*, N.Y. TIMES, Dec. 13, 2007

DONALD C. SHOUP, THE HIGH COST OF FREE PARKING (2005)

Bruce Siceloff, *Drivers Might Pay Road Taxes by Mile*, NEWS OBSERVER, Jun. 17, 2007, available at http:www.newsobserver.com/news/growth/v-print/story/607113.html

PETER SINGER & JIM MASON, THE WAY WE EAT: WHY OUR CHOICES MATTER (2006)

S. FRED SINGER & DENNIS T. AVERY, UNSTOPPABLE GLOBAL WARMING: EVERY 1,500 YEARS (2007)

J.B. SMITH & D.A. TIRPAK, THE POTENTIAL EFFECTS OF GLOBAL CLIMATE CHANGE ON THE UNITED STATES (1990)

PETER F. SMITH, ARCHITECTURE IN A CLIMATE OF CHANGE: A GUIDE FOR SUSTAINABLE DESIGN (2001)

Claudia Dias Soares, *Critical Issues in Implementing Energy Taxation*, in EU CLIMATE CHANGE POLICY: THE CHALLENGE OF NEW REGULATORY INITIATIVES 256 (Marjan Peeters & Kurt Deketelaere eds. 2006)

GEORGE SOROS ON GLOBALIZATION (2002)

Steven Sorrell & Jos Sijm, *Carbon Trading in the Policy* Mix, in CLIMATE-CHANGE POLICY 194 (Dieter Helm ed.2005)

Mark J. Spalding & Charlotte de Fontaubert, *Conflict Resolution for Addressing Climate Change With Ocean-Altering Projects*, 37 ENV'T L. RPTR. 10740 (2007)

Bradford C. Spencer, Note, *Evaluating Kentucky's Investment Tax Credits in Light of Cuno v. DaimlerChrysler, Inc.*, 94 KY. L.J. 161 (2005–2006)

JAMES GUSTAVE SPETH, RED SKY AT MORNING: AMERICA AND THE CRISIS OF THE GLOBAL ENVIRONMENT (2004)

Robert Stacey, *Urban Growth Boundaries: Saying "Yes" to Strengthening Communities*, 34 CONN. L. REV. 597 (2002)

JAMES STEELE, ECOLOGICAL ARCHITECTURE: A CRITICAL HISTORY (2005)

Eva Steele-Saccio, *Education by Design*, GOOD MAGAZINE, available at http://www.goodmagazine.com/section/Features/education-by-design

Eleanor Stein, *Regional Initiatives to Reduce Greenhouse Gas Emissions*, in GLOBAL CLIMATE CHANGE AND U.S. LAW 315 (Michael B. Gerrard ed. 2007)

NICHOLAS STERN, STERN REVIEW: THE ECONOMICS OF CLIMATE CHANGE (2007)

Joseph E. Stiglitz & Andrew Charlton, Fair Trade For All: How Trade Can Promote Development (2005)
Strategic Planning of Sustainable Urban Water Management (Per-Arne Malmqvist et al., eds. 2006)
Howard E. Susman & Kathleen J. Doll, *Wind Advisory: Finding a Suitable Site for a Wind Farm Requires More Than Locating a Blustery Location*, 30 Los Angeles Law. 35 (Jan. 2008)
Edna Sussman, *Reshaping Municipal and County Laws to Foster Green Building, Energy Efficiency, and Renewable Energy*, 16 N.Y.U. Envtl. L.J. 1 (2008)
Terry Tamminen, Lives Per Gallon: The True Cost of Our Oil Addiction (2006)
A. Dan Tarlock & Sarah B. Van De Wetering, *Water and Western Growth*, 59 Planning & Envt'l L. 3 (No. 5 May 2007)
Technology: Don't Call it a Blimp, 210 Nat'l Geographic, Nov. 2006, at 14
Peter Tertzakian, A Thousand Barrels a Second: The Coming Oil Break Point and the Challenges Facing an Energy Dependent World (2006)
Time to Tax Carbon, L.A. Times, May 28, 2007
Patrick E. Tolan, Jr., *Tax and Insurance Consequences of Major Disasters: Weathering the Storm*, 31 Nova L. Rev. 487 (2007)
Top 10 Things You Can Do to Reduce Global Warming, available at http://environment.about.com/od/globalwarming/tp/globalwarmtips.htm
Top Tips to Stop Global Warming, aol.com, Sept. 2, 2007, available at http://reference.aol.com/planet-earth/global-warming/top-tips-stop-global-warming
Track I, Annex C, International Strategy for Disaster Reduction (2005)
Track II, Annex D, International Strategy for Disaster Reduction (2005)
Sarah M. Tran, *Updated Hurricane Models: A New Opportunity to Insure Against Climate Change*, 14 B.U. J. Sci. & Tech. L. 73 (2008)
Transportation Finance at the Ballot Box: Voters Support Increased Investment & Choice (2007), available at http://www.cfte.org/success/TrendsReport00-05.asp
Rebecca Tsosie, *Indigenous People and Environmental Justice: The Impact of Climate Change*, 78 U. Colo. L. Rev. 1625 (2007)
Eleanor G. Turman, *Regional Impact Assessments: A Case Study of California*, in Climate Change Policy: A Survey 89 (Stephen H. Schneider, Armin Rosencranz & John O. Niles eds. 2002)
Chris Turner, *The Secret Greening of Calgary*, Globe & Mail, Sept. 15, 2007

Twenty in Ten: Strengthening America's Energy Security, *available at* http://www.whitehouse.gov/stateoftheunion/2007/initiatives/energy.html

United Nations Environment Programme, GEO4—Global Environment Outlook (2007)

U.S. Department of Agriculture, *How Anaerobic Digestion (Methane Recovery) Works*, *available at* http://www.eere.energy.gov/consumer/your_workplace/farms_ranches/index.cfm/mytopic=300

U.S. Dep't of Energy, Energy Info. Admin., Annual Energy Outlook 2006 with Projections to 2030 (2006), *available at* http://www.eia.doe.gov/oiaf/aeo/index.html

U.S. Environmental Protection Agency, ENERGY STAR, Product Specifications, Eligibility Criteria & Partner Commitments, *available at* http://www.energystar.gov/index.cfm?c=product_specs.pt_product_specs (last visited Mar. 24, 2006)

U.S. Environmental Protection Agency, Global Warming, Actions, State, http://yosemite.epa.gov/OAR/globalwarming.nsf/content/ActionsState.html

U.S. Loans for Coal Plants Clash with Carbon Cuts, Wash. Post.com, May 14, 2007

Brenda Vale & Robert Vale, Green Architecture: Design for an Energy-Conscious Future (1991)

Michael P. Vandenbergh, & Anne C. Steinemann, *The Carbon-Neutral Individual*, 82 N.Y.U. L. Rev. 1673 (2007)

Mary Lynne Vellinga, *Aspiring to Be America's Greenest City: State Commitment Helps Sacramento Rank No. 2 in Energy Efficient Office Space*, Sacramento Bee, Sept. 5, 2007, available at http:www.sacbee.com/378/v-print/story/360985.html

Verticle Farming, Wikipedia, *available at* http://en.wikipedia.org/wiki/Vertical_farming

Carol Hardy Vincent et al., Federal Land Management Agencies: Background on Land and Resources Management, in Samuel T. Prescott, Federal Land Management: Current Issues and Background 37, 37 (2003)

Gretchen Vogel, *Upending the Traditional Farm: Cities are Taking Over Farmland. Could They Someday Take Over the Job of Farming, Too*, 319 Science 752 (Feb. 2008)

Frederic H. Wagner, *Global Warming Effects on Climactically-Imposed Ecological Gradients in the West*, 27 J. Land Resources & Envtl. L. 109 (2007)

Matthew L. Wald, *Study How U.S. Could Cut 28% of Greenhouse Gases*, N.Y. Times, Nov. 30, 2007)

Michael B. Walsh, Comment, *A Rising Tide in Renewable Energy: The Future of Tidal In-Stream Energy Conversion (TISEC)*, 19 VILL. ENVTL. L.J. 193 (2008)

DIANE RAINES WARD, WATER WARS: DROUGHT, FLOOD, FOLLY, AND THE POLITICS OF THIRST (2002)

PETER D. WARD, UNDER A GREEN SKY: GLOBAL WARMING, THE MASS EXTINCTIONS OF THE PAST AND WHAT THEY CAN TELL US ABOUT OUR FUTURE (2007)

Rick Wartzman, *Can the City Save the Farm?*, CAL. MAGAZINE, May/June, 2007, available at http://www.newamerica.net/publications/articles/2007/can_city_save_farm_5422

WATERS IN PERIL (Leah Bendell-Young & Patricia Gallaugher eds. 2001)

WATER IN CRISIS: A GUIDE TO THE WORLD'S FRESH WATER RESOURCES (H. Peter Gleich ed. 1993)

Ryan Waterman, Comment, *Addressing California's Uncertain Water Future by Coordinating Long-Term Land Use and Water Planning: Is a Water Element in the General Plan the Next Step?*, 31 ECOLOGY L.Q. 117 (2004)

SPENCER R. WEART, THE DISCOVERY OF GLOBAL WARMING (2003)

Ann Brewster Weeks, *Subseabed Carbon Dioxide Sequestration as a Climate Mitigation Option for the Eastern United States: A Preliminary Assessment of Technology and Law*, 12 OCEAN & COASTAL L.J. 245 (2007)

Rachel Weinberger, *The High Cost of Free Highways*, 43 IDAHO L. REV. 475 (2007)

Mitch Weiss, *Drought Could Shut Down Nuclear Plants*, ASSOCIATED PRESS, Jan. 23, 2008

Patricia Weisselberg, Comment, *Shaping the Energy Future in the American West: Can California Curb Greenhouse Gas Emissions from Out-of-State, Coal-Fired Power Plants Without Violating the Dormant Commerce Clause?*, 42 U.S.F.L. REV. 185 (2007)

WET GROWTH: SHOULD WATER LAW CONTROL LAND USE? (Craig Anthony (Tony) Arnold ed. 2005)

Timothy B. Wheeler, *Searching for Signs of Intelligent Growth*, HARTFORD COURANT, Sept. 2, 2007

Hope Whitney, *Cities and Superfund: Encouraging Brownfield Redevelopment*, 30 ECOLOGY L.Q. 59 (2003)

WILLIAM H. WHYTE, THE LAST LANDSCAPE (2002)

Jonathan Baert Wiener, *Designing Global Climate Regulation*, in CLIMATE CHANGE POLICY: A SURVEY 151 (Stephen H. Schneider, Armin Rosencranz & John O. Niles eds. 2002)

Jonathan B. Wiener, *Think Globally, Act Globally: The Limits of Local Climate Policies*, 155 U. PA. L. REV. 1961 (2007)

George F. Will, *Fuzzy Climate Math*, WASH. POST, Apr. 12, 2007, at A27, available at http://www.skepticism.net/faq/environment/global_warming/index.html

CHARLES WOHLFORTH, THE WHALE AND THE SUPERCOMPUTER: ON THE NORTHERN FRONT OF CLIMATE CHANGE (2004)

Richard Wolfson & Stephen H. Schneider, *Understanding Climate Science, in* CLIMATE CHANGE POLICY: A SURVEY (Stephen H. Schneider, Armin Rosencranz & John O. Niles eds. 2002)

Martin Wolk, *Cost of War Could Surpass $1 Trillion*, MSNBC, Mar. 17, 2006, available at http://www.msnbc.msn.com/id/11880954/

David Wortman, *No Such Thing as a Free Ride*, SUSTAINABLE INDUSTRIES J., Aug. 3, 2007, available at http://www.printthis.clickability.com/pt/cpt?acrtion=Sustainable+Industries+Journal

Matt Wrye, *State Drops its Lawsuit*, SAN BERNARDINO COUNTY SUN, Aug. 22, 2007

Nicholle Winters, *Carbon Dioxide: A Pollutant in the Air, But is the EPA Correct that It Is Not an "Air Pollutant"?*, 104 COLUM. L. REV. 1996 (2004)

Johan Woltjer & Niels Al, *Integrating Water Management and Spatial Planning*, 73 J. AM. PLANNING ASS'N 211 (2007)

Mary Christina Wood, *Nature's Trust: A Legal, Political, and Moral Frame for Global Warming*, 34 B.C. ENVTL. AFF. L. REV. 577 (2007)

William Yardley, *Victim of Climate Change, a Town Seeks a Lifeline*, N.Y. TIMES, May 27, 2007

http://yosemite.epa.gov/OAR/globalwarming.nsf/UniqueKeyLookup/SHSU5BUTYL/$File/netmetering.pdf

Samantha Young, *Ca. Land-Use Dispute Complicates Budget*, FORBES.COM, July 26, 2007, available at http://www.forbes.com/feeds/ap/2007/07/26/ap3956150.html

Kamaal R. Zaidi, *High Speed Rail Transit: Developing the Case for Alternative Transportation Schemes in the Context of Innovative and Sustainable Global Transportation Law and Policy*, 26 TEMP. J. SCI. TECH. & ENVTL. L. 301 (2007)

Alan Zarembo, *Can You Buy a Greener Conscience?: A Budding Industry Sells 'Offsets' of Carbon Emissions, Investing in Environmental Projects. But There Are Doubts About Whether it Works*, L.A. TIMES, Sept. 2, 2007, available at http://www.latimes.com/news/science/environmental/la-sci-offsets2sep02,1,5021370.story?ctrack=1&cset=true

Matthew D. Zinn, *Adapting to Climate Change: Environmental Law in a Warmer World*, 34 ECOLOGY L.Q. 61 (2007)

Table of Statutes, Constitutional Provisions, Regulations, and Executive Orders

U.S. Const., Art. I, §8, 29, 104, 188
U.S. Const., Art. IV, §2, 29
American Jobs Creation Act of 2004, Pub. L. No. 108-357, 118 Stat. 1418 (2004), 26 U.S.C.A. 896–907 (2007), 176
Clean Coal Power Initiative Act, 42 U.S.C.A. §§15961–64 (2007), 95
Comprehensive Environmental Response, Compensation, and Liability Act of 1980 (CERCLA), 42 U.S.C. §§9601–9675 (2000), 38
Disaster Mitigation Act of 2000, Pub. L. No. 106-390, §322 (2000), codified at 42 U.S.C. §5165(a); 44 C.F.R. §201 (2002), 78
Endangered Species Act of 1973, 16 U.S.C. §§1531–44 (2000), 153, 156
Energy Policy Act of 2005, §1501, 42 U.S.C.A. §7545(o)(2)(B)(I) (2007), 112
Energy Policy and Conservation Act, 49 U.S.C. §§32901–19 (2000), 191
National Appliance Energy Conservation Act of 1987, Pub. L. No. 100-12, 101 Stat. 103, *codified at* 42 U.S.C. §§6291–97, 6299, 6302, 6303, 6305–6, 6308–9 (2000), 56
Post-Katrina Emergency Management Reform Act of 2006, Pub. L. No. 109-295, tit. VI, 120 Stat. 1394 (2006), 78-80, 83, 85
Safe, Accountable, Flexible, Efficient Transportation Equity Act: A Legacy for Users (SAFETEA-LU), Pub. L. No. 109-59, 119 Stat. 1144 (2005), *codified at* 23 U.S.C.A. §134(h)(1)(C) (West 2006), 189-90
Alaska Stat. §§45.88.010 to 45.88.140 (2006), 108
Ariz. Exec. Order No. 2005-05 (2005), *available at* http://www.governor.state.az.us/eo/2005_05.pdf, 109

ARIZ. REV. STAT. § 41-1510.01 (2006), 108
ARIZ. REV. STAT. § 42-5061 (Supp. 2007), 109
ARIZ. REV. STAT. § 42-5075(B)(14) (Supp. 2007), 109
ARIZ. REV. STAT. § 43-1083 (Supp. 2007), 108
ARIZ. REV. STAT. ANN. § 49-542 (Supp. 2007), 106
ARIZ. REV. STAT. ANN. §§ 1375 to 1375.03 (Supp. 2007), 110
ARK. CODE ANN. § 15-4-2104 (2003), 108
ARK. CODE ANN. §§ 15-4-2801 to 15-4-2805 (2003), 108
ARK. CODE ANN. § 15-4-2804 (2003), 107
ARK. CODE ANN. §§ 23-18-601 to 23-18-604 (2002 & Supp. 2007), 107
CAL. ADMIN. CODE §§ 1601 to 1608 (2008), 110
13 CAL. CODE REGS. § 1961.1 (2008), 106
California Energy Comm'n, Appliance Efficiency & Appliance Regulations http://www.energy.ca.gov/efficiency/appliances/, 110
Cal. Exec. Order No. S-20-04 (2004), *available at* http://www.dot.ca.gov/hq/energy/ ExecOrderS-20-04.htm, 109
California Global Warming Solutions Act of 2006, *available at* http://www.leginfo.ca.gov/cgi-bin/postquery?bill_number=ab_32&sess= PREV&*house=B&author=nunez,* 104
CAL. GOV'T CODE § 14684 (West Supp. 2008), 109
CAL. GOV. CODE § 66473.7(a)(2) (2003), 199
CAL. HEALTH & SAFETY CODE §§ 38500–38599 (West Supp. 2008) ("Global Warming Solutions Act of 2006," Stats. 2006, c. 488 (A.B. 32); 106, 149, 180
CAL. HEALTH & SAFETY CODE § 38562 (West Supp. 2008) (A.B. 32), 106
CAL. HEALTH & SAFETY CODE § 43018.5 (West 2006 & Supp. 2008), 106-7
CAL. PUB. RES. CODE §§ 25740–25751 (West 2007 & Supp. 2008), 107
CAL. PUB. RES. CODE § 25751 (West 2006), 106
CAL. PUB. UTIL. CODE § 399.11 (West Supp. 2008), 105
CAL. PUB. UTIL. CODE § 2827 (West Supp. 2008), 107
CAL. REV. & TAX CODE § 73, 17208.1 (West Supp. 2008), 109
Cal. S.B. 1368, Act of Sept. 29, 2006, ch. 598, § 2, 2006 Cal. Legis. Serv. at 3795, 104
CAL. PUB. UTIL. CODE § 399.11 (West Supp. 2008), 105
4 COLO. CODE REGS. § 723-3, Rule 3664 (2007), 107
COLO. REV. STAT. ANN. § 39-22-516 (2007), 108
COLO. REV. STAT. ANN. § 40-2-124 (West 2007), 105
COLO. REV. STAT. ANN. §§ 40-9.5-301 to 40-9.5-305 (West 2007), 107
CONN. AGENCIES REGS. § 22a-174-36b (2005), 106

CONN. GEN. STAT. § 16a-27 (West 2007), 151
CONN. GEN. STAT. § 16a-48 (West 2007), 110
CONN. GEN. STAT. § 16-245a (West 2007) 105
DEL. CODE ANN. tit. 26, § 354 (Supp. 2006), 105
D.C. CODE Ch. 14A §§ 6-1451.01 to 6-1451.11 (Supp. 2007) (Green Building Act of 2006), 121
FLA. STAT. ANN. § 212.08 (West Supp. 2008), 109
FLA. STAT. ANN. § 220.192 (West Supp. 2008), 108
FLA. STAT. ANN. § 220.193 (West Supp. 2008), 108
FLA. STAT. ANN. § 316.0741 (West 2006), 111
FLA. STAT. ANN. §§ 377.805 to 377.806 (West Supp. 2008), 107
GA. CODE ANN. § 32-9-4 (2006), 111
GA. CODE ANN. §§ 46-3-50 to 46-3-56 (2004), 107
GA. CODE ANN. §§ 48-7-40.16 (2005), 108
HAW. REV. STAT. § 46-19.6 (Supp. 2006), 110, 111
HAW. REV. STAT. § 103D-412 (Supp. 2006), 107
HAW. REV. STAT. ANN. § 196-7 (Supp. 2006), 110
HAW. REV. STAT. § 196-9 (Supp. 2006), 110, 111
HAW. REV. STAT. ANN. § 235-12.5 (Supp. 2006), 108
HAW. REV. STAT. ANN. §§ 269-92 (Supp. 2006), 105
HAW. REV. STAT. ANN. §§ 269-101 to 269-111 (Supp. 2006), 107
HAW. REV. STAT. § 269-121 to 269-124 (Supp. 2006), 107
2006 Haw. Sess. Laws 096, Pt. 2, §2, *available at* http://www.capitol.hawaii.gov/session2006/bills/HB2175_cd1_.htm, 107
IDAHO CODE ANN. § 39-124(Supp. 2007), 106
IDAHO CODE ANN. § 39-125 (Supp. 2007), 106
IDAHO CODE ANN. § 39-4109 (Supp. 2007), 110
IDAHO CODE ANN. § 39-4116 (Supp. 2007), 110
170 IND. ADMIN. CODE R. 4-4.2 (2006), 107
20 ILL. COMP. STAT. ANN. 687 (West Supp. 2007), 105
20 ILL. COMP. STAT. ANN. 687/6-3(West Supp. 2007), 107
20 ILL. COMP. STAT. ANN. 687/6-4 (West Supp. 2007), 107, 108
20 ILL. COMP. STAT. ANN. 3105/10.04 (West Supp. 2007), 109
20 ILL. COMP. STAT. 3953/25 (West 2007), 110
220 ILL. COMP. STAT. ANN. 5/16-111.1 (West 2007), 107
415 ILL. COMP. STAT. 145/1 (West 2008), 106
170 Ind. Admin. Code R. 4-4.2 (2008), 107
IOWA CODE ANN. § 266.39C (West 2003), 107
IOWA CODE ANN. § 423.3 (West Supp. 2007), 109

Iowa Code Ann. § 437A.6 (West 2006), 109
Iowa Code Ann. § 476.46 (West Supp. 2007), 108
Iowa Code Ann. § 476C (West Supp. 2007), 108
Iowa Exec. Order No. 41 (2005), *available at* http://www.dsireusa.org/documents/ Incentives/IA08R.pdf, 107, 109, 111
Kan. Stat. Ann. § 66-1,184 (2002), 107
Kan. Stat. Ann. § 79-201h (1997), 109
Ky. Rev. Stat. Ann. §§ 278.465 to 278.468 (Supp. 2007), 107
La. Rev. Stat. Ann. § 47:38 (2001), 108
La. Rev. Stat. Ann. § 47:287.757 (2001), 108
La. Rev. Stat. Ann. § 47:1706 (2006), 109
Me. Code R. § 06-096, ch. 127 § 2 (2008), 106
Me. Rev. Stat. Ann. tit. 5, § 1812E (West 2002), 111
Me. Rev. Stat. Ann. tit. 29-A, § 102 (Supp. 2007), 111
Me. Rev. Stat. Ann. tit. 35-A, § 3209(3) (2007), 109
Me. Rev. Stat. Ann. 35A § 3210(3)-A (2007), 105
Me. Rev. Stat. Ann tit. 35-A. § 3211-C (Supp. 2007), 107
Me. Rev. Stat. Ann. tit. 36, § 5219-P (Supp. 2007), 108
Me. Rev. Stat. Ann. tit. 36, § 5219-AA (Supp. 2007), 108
Md. Code Ann., Pub. Util. Cos. § 7-306 (Supp. 2007), 107
Md. Code Ann., Pub. Util. Cos. § 7-703 (Supp. 2007), 105
Md. Code Ann., State Gov't § 9-2007 (Supp. 2007), 107
Md. Code Ann., Tax-Gen. § 10-722 (Supp. 2004), 108
Md. Code Ann., Tax-Gen. § 11-207 (2004), 109
225 Mass Code Regs. 14.07 (2007), 105
310 Mass. Code Regs. 7.29 (2008), 106
310 Mass. Code Regs. 7.40 (2005), 106
Mass. Gen. Laws ch. 25A, § 11F (West 2002), 105
Mass. Gen. Laws Ann. Ch. 40R §§ 1–14 (West. Supp. 2007), 151
Mich. Comp. Laws Ann. § 208.39e (West 2003), 108
Mich. Comp. Laws Ann. § 211.9i (West 2005), 109
Minn. Stat. Ann. § 16B.32 (West 2005), 109
Minn. Stat. Ann. § 41B.046 (West Supp. 2008), 108
Minn. Stat. Ann. § 216B.241 (West Supp. 2007), 107-8
Minn. Stat. Ann. § 216B.1691 (West 2006), 105
Minn. Stat. Ann. § 216C.41 (West Supp. 2008), 107-8
Minn. Stat. Ann. § 272.02 (West 2007), 109
Minn. Stat. Ann. § 272.028 (West 2007), 109
Minn. Stat. Ann. § 297A.67 (West 2007), 109

Minn. Stat. Ann. §297A.68 (West 2007), 109
Miss. Code Ann. §57-39-39 (West 1999), 108
Mo. Ann. Stat. §§135.300 to 135.311 (West 2000), 108
Mo. Ann. Stat. §386.887 (West 2007), 107
Mo. Ann. Stat. §§640.651 to 640.686 (West 2006), 108
Mo. Code Regs. Ann. tit. 4, §240-20.065 (2007), 107
Mont. Code Ann. §15-6-225 (2007), 109
Mont. Code Ann. §15-24-1401 (2007), 109
Mont. Code Ann. §15-32-115 (2007), 108
Mont. Code Ann. §15-32-401 (2007), 108
Mont. Code Ann. §15-72-104 (2007), 109
Mont. Code Ann. §§69-8-601 to 69-8-605 (2007), 107
Mont. Code Ann. §69-3-2004 (2007), 105
Mont. Code Ann. §75-25-101 (2007), 108
Mont. Code Ann. §80-12-201 (2007), 108
Neb. Rev. Stat. §§66-1001 to 66-1011 (2003), 108
Nev. Rev. Stat. §111.239 (Nexis Lexis Supp. 2005), 110
Nev. Rev. Stat. §278.0208 (Nexis Lexis Supp. 2005), 110
Nev. Rev. Stat. §278.580 (Nexis Lexis 2002 & Supp. 2005), 110
Nev. Rev. Stat. §361.0687 (Nexis Lexis 2007), 109
Nev. Rev. Stat. §704.768 (Nexis Lexis 2005), 107
Nev. Rev. Stat. Ann §704.7821 (LexisNexis Supp. 2005), 105
N.H. Rev. Stat. Ann. §§125-0:3 to 0:4 (Supp. 2007), 106
N.H. Rev. Stat. §362-A:9 (Supp. 2007), 107
N.H. Rev. Stat. Ann. §362-F:10 (West Supp. 2007), 108
N.H. Rev. Stat. Ann. §§477:49 to 477:51 (2001), 110
N.H. Rev. Stat. Ann. §3211-C (2006), 108
N.J. Rev. Stat. §§48:3-99 to 48:3-106 (2005), 110
7 N.J. Admin. Code §27B-5.3 (2008), 106
13 N.J. Admin. Code §20-43.1 (2008), 106
13 N.J. Admin. Code §43.21 (2008), 106
N.J. Admin. Code tit. 14, §8-2 (2006), 106
N.J. Admin. Code tit. 14, §8-2.3 (2008), 106
N.M. Admin. Code tit. 17, §9.572 (2005), 106
N.M. Stat. Ann. §§6-21D-1 to 6-21D-10 (2005), 108
N.M. Stat. Ann. §7-2A-19 (2007), 108
N.M. Stat. §7-9G-2 (2007), 108
N.M. Stat. Ann. §§47-3-1 to 47-3-5 (1995), 110
N.M. Stat. Ann. §§68-2-29 to 68-2-33 (2002), 108

N.M. Stat. Ann. §71-7-6 (2007), 108
N.Y. Comp. Code R. & Regs. tit. 6, §§210, 218 (2005), 106
N.Y. Energy Law Appx. Pt. 7825.1 to 7826.1 (McKinney 2004), 110
N.D. Cent. Code §54-44.5-09 (Supp. 2007), 108
N.D. Cent. Code §57-02-08(27) (Supp. 2007), 109
N.D. Cent. Code §57-38-01.8 (Supp. 2007), 108
N.D. Cent. Code §§57-39.2-04 to 57-43.2-03 (Supp. 2007), 109
Ohio Rev. Code Ann. §§4928.61 to 4928.63 (Supp. 2007), 108
Ohio Rev. Code Ann. §4928.67 (West 2000), 107
Okla. Stat. Ann. tit. 68, §2357.32A (West Supp. 2008), 108
Or. Rev. Stat. §90.265 (2007), 110
Or. Rev. Stat. §307.175 (2007), 109
Or. Rev. Stat. §§316.116, 469.165 to 469.170 (2005), 108
Or. Rev. Stat. §468A.365 (2005), 111
Or. Rev. Stat. §469.501 (2005), 106
Or. Rev. Stat. §§470.050 to 470.210 (2005), 108
Or. Rev. Stat. §757.300 (2005), 107
Or. Rev. Stat. §803.350(8)(a) (2007), 106
Or. Rev. Stat. §§815.295 to 815.300 (2005), 111
Or. Admin R. 345-024-0550 (2007), 106
Or. Admin R. 345-024-0560 (2008), 106
73 Pa. Cons. Stat. §1647.3 (West. Supp. 2007), 108
73 Pa. Cons. Stat. §§1648.1–.8 (West 2004), 106
73 Pa. Cons. Stat. §1648.2 (West 2007), 107
73 Pa. Cons. Stat. §1648.5 (West 2007), 107
75 Pa. Cons. Stat. Ann. §4706(c) (West 2006), 106
R.I. Gen. Laws §§31-47.1-1 to 47.1-11 (2002), 106
R.I. Gen. Laws §39-2-1.2(b) (2006), 108
R.I. Gen. Laws §39-26-4 (2006), 106
R.I. Gen. Laws §§39-27-1 to 33-27-9 (2006), 110
R.I. Gen. Laws §44-3-21 (2005), 109
R.I. Gen. Laws §44-18-40.1 (2005), 109
R.I. Gen. Laws §§44-57.1 to 44-57.12 (2005), 109
S.C Code Ann. §12-6-3587 (2007), 109
S.C Code Ann. §12-36-2110 (2000 & Supp. 2007), 109
S.C. Exec. Order No. 2001-35 (2001), *available at* http://www.scstatehouse.net/archives/executiveorders/exor0135.htm, 107
S.D. Codified Law §§10-6-35.8 to 10-6-35.20, 10-4-36 to 38 (2004), 109
Tenn. Code Ann. §4-3-710 (2005), 108

TENN. CODE ANN. §67-5-601 (2006), 109
TEX. GOV'T CODE ANN. §2166.403 (Vernon Supp. 2007), 110
TEX. UTIL CODE ANN. §39.904 (Vernon 2005), 106
UTAH CODE ANN. §54-15-102 (Supp. 2007), 107
UTAH CODE ANN. §54-15-103 (Supp. 2007), 107
UTAH CODE ANN. §59-10-1009 (Supp. 2007), 109
UTAH CODE ANN. §59-12-104 (2006), 109
VT. CODE R. §12-031-001 (2007), 106
Vt. Exec. Order No. 14-03 (2003), *available at* http://governor.vermont.gov/tools/index.php?topic=ExecutiveOrders&id=249&v=Article, 110
VT. STAT. ANN. §§9-2791 to 9-2798 (2006), 110
VT. STAT. ANN. tit. 30, §219a (Supp. 2007), 107
VT. STAT. ANN. tit. 32, §9741(46) (2001 & Supp. 2007), 109
VA. CODE ANN. §45.1-392 (2002), 108
WASH. REV. CODE ANN. §§19.260.010 to 19.260.900 (West 2007), 110
WASH. REV. CODE ANN. §70.94.960 (West 2002), 108
WASH. REV. CODE ANN. §§70.120A.010 (West Supp. 2008), 106
WASH. REV. CODE ANN. §§80.60.005 to 80.60.040 (West 2001 & Supp. 2008), 107
WASH. REV. CODE ANN. §80.70.20 (West Supp. 2008), 106
W. VA. CODE ANN. §11-6A-5a (LexisNexis 2003 & Supp. 2007), 109
Wis. Exec. Order No. 145 (2006), *available at* http://www.wisgov.state.wi.us/journal_media_detail.asp?prid=1907, 110, 111
WIS. STAT. ANN. §13.48 (West Supp. 2007), 110
WIS. STAT. ANN. §16.75 (West Supp. 2007), 111
WIS. STAT. ANN. §16.957 (West 2003), 107
WIS. STAT. ANN. §70.111(18) (West Supp. 2005), 109
WIS. STAT. ANN. §101.027 (West Supp. 2007), 110
WIS. STAT. ANN. §196.378 (West 2005), 106
WYO. STAT. ANN. §39-15-105 (a)(viii)(N) (2007), 109
WYO. STAT. ANN. §§37-16-101 to 37-16-104 (2007), 107

Table of Cases

Center for Biological Diversity v. National Highway Traffic Safety Admin., 508 F.3d 508 (9th Cir. 2007), 191

Central Valley Chrysler Valley Jeep, Inc., v. Witherspoon, No.CV-F-04-6663 AWI LJO, 2006 WL 2473663 (E.D. Cal. Aug. 25, 2006), 30

City of Philadelphia v. New Jersey, 437 U.S. 617 (1978), 29

Clark v. Allen, 331 U.S. 503 (1947), 188

Farm Raised Salmon Cases, *In re*, 175 P.3d 1170 (Cal. 2008), 155

Green Mountain Chrysler Plymouth Dodge Jeep v. Crombie, 508 F. Supp. 2d 295 (D. Vt. 2007), 187

H.P. Hood & Sons, Inc. v. DuMond, 336 U.S. 529 (1939), 29

In re Katrina Canal Breaches Consolidated Litigation, 471 F. Supp. 2d 684 (E.D. La. 2007), 200

Maine v. Taylor, 477 U.S. 131 (1986), 29

Massachusetts v. EPA, 127 S. Ct. 1438 (2007), 8, 191

N.Y. Pub. Serv. Comm'n, Case No. 03-E-0188 (Sept. 24, 2004), *available at* http://www3.dps.state.ny.us/pscweb/WebFileRoom.nsf/0/85D8CCC6A42DB86F85256F1900533518/$File/301.03e0188.RPS.pdf?OpenElement, 106

San Antonio Indep. School Dist. v. Rodriguez, 411 U.S. 1 (1973), 69

Sheff v. O'Neill, 678 A.2d 1267 (Conn. 1996), 69-70

Supreme Court of New Hampshire v. Piper, 470 U.S. 274 (1985), 29

United Haulers Ass'n v. Oneida-Herkimer Solid Waste Management Auth., 127 S. Ct. 1786 (2007), 29

West Lynn Creamery, Inc. v. Healy, 512 U.S. 186 (1994), 29

Zelman v. Simmons-Harris, 536 U.S. 639 (2002), 69

Zschernig v. Miller, 389 U.S. 429 (1968), 188

Index

Abatzoglou *et al.*, *A Primer on Global Climate Change and Its Likely Impacts*, 15
Abortion, 139
Abrams, *Invisible Hitchhikers: Slide Behind the Wheel and your Biggest Health Threat May Not be the Other Drivers On the Road, but the Invisible Toxins Riding Shotgun*, 163
Adams, Comment, *Bringing Green Power to the Public Lands: The Bureau of Land Management's Authority and Discretion to Regulate Wind-Energy Developments*, 111
Adams *et al.*, *Impacts on the Agricultural Sector*, 28
Adler, *Corrective Justice and Liability for Global Warming*, 150
Aforestation, 28
Africa, 42, 136, 195
Agriculture, ch2
    Degradation, 25-26
    Farmers' Markets, 27
    Fertilizer, 25
    Herbicides, 25-26
    Irrigation, 24
    Drip irrigation, 47
    Manure digesters, 28
    Methane, 28
    Pesticides, 25, 27
    Subsidies, 31-32
    Topsoil, 25
    Urban high rise farms, 20
Alaska, 17
Alberta, Canada, 177
Albian Sands, Canada, 177
Alexandria, Virginia, 142

Allen *et al.*, *Scientific Challenges in the Attribution of Harm to Human Influence on Climate*, 9
Altamont, California, 99
AMERICAN SOCIETY OF LANDSCAPE ARCHITECTS, THE SUSTAINABLE SITES INITIATIVE—STANDARDS AND GUIDELINES: PRELIMINARY REPORT, n627
Amsterdam, Netherlands, 116
ANI, *Family Planning Rules in China Tightened*, 139
Antarctica, 16
Appliances, 43
    Air conditioning, 43, 101
    Auto switching power strips, 48
    Clothes dryers, 44
    Computers, 48-49
    Garden tools, 52
    Hot water, 48
    Laser printers, 49
    Microwave ovens, 44-45
    Ovens, 44
    Refrigerator, 43
    Televisions, 48
    Unplugging, 48
    Washing machines, 43-44
Arnold, *Is Wet Growth Smarter Than Smart Growth?: The Fragmentation and Integration of Land Use and Water*, 199
Architecture
    Air conditioning, 43, 44, 101
    Curtains, 44
    Fireplaces, 44
    Green, 53-54
    Heating, 44, 101
    Plumbing, 47
    Roof Gardens, 27

Architecture, *continued*
  Swimming pool, 47
  Thermostat, 44
Arctic, 16-17
Arctic Basin, 153
Arctic National Wildlife Refuge (ANWR), 89
Army Corps of Engineers, 200
Arup to Design Infrastructure for Personal Rapid Transport System at Heathrow, 186
Asia, 25-26, 135-36, 195
Associated Press, *Climate Report: 'Highway to Extinction:' Dire Predictions Includes Loss of Species, Increasing Scarcity of Water*, 15, 16
Associated Press, *Climate Report Sound Dire Warnings: Global Warming Could Mean Hundreds of Millions without Water*, 10-11, 15, 16, 18
Associated Press, *Study: Corn Boom Could Expand 'Dead Zone' in Gulf*, 24, 100, 175
Associated Press, *Most Polar Bears Could Die Out by 2050: U.S. Geological Survey Says Two-Thirds Could Vanish Because of Ebbing Ice*, 153, 154
Associated Press, *Study: Corn Ethanol No Climate Solution: Greenhouse Emissions Much Higher if Land Use Factored In, Researchers Say*, 100
Atlanta, 16, 144
Atlantic City, 17
Attenborough, Sir David, 203
Australia, 32
  Water use, 47
Autofreie Siedlung, Munster, Germany, 119
Automobiles, 49, 51-52
  Air Pollution, 173
  China, 165
  Costs, 162
  Health, 163
  Hybrids, 51, 52-53, 167, 179-80
  India, 165
  Number, 161
  Ownership, 165
  Parking, 168-69
  Subsidies, 166-67
  Vehicle Miles Traveled, 161, 162-63
Ayers, *The Energy We Overlook*, WORLDWATCH, 90
Babbitt, *The Case for Conditioning Federal Infrastructure Investment on State Land and Water Planning*, 201
Backes & Teuben, *Legal Aspects of the Dutch Approach to CO_2*, 52
Baghdad, 16
Bailey, Comment, *The Sink that Sank the Hague: A Comment on the Kyoto Protocol*, 112
Bangkok, 17
BANISTER, UNSUSTAINABLE TRANSPORT: CITY TRANSPORT IN THE NEW CENTURY, 30, 165, 179, 185
Barnett & Hill, *Design for Rising Sea Levels*, 17, 80
Bartholomew, *The Machine, the Garden, and the City: Toward an Access-Efficient Transportation Planning System*, 190
BARTSCH & COLLATON, BROWNFIELDS: CLEANING AND REUSING CONTAMINATED PROPERTIES, 35
Bea et al., *Federal Emergency Management Policy Changes After Hurricane Katrina: A Summary of Statutory Provisions*, 79, 80
BEATLEY, GREEN URBANISM: LEARNING FROM EUROPEAN CITIES, 30, 54, 122, 185
Beatley & Collins, *Smart Growth and Beyond: Transitioning to a Sustainable Society*, 142
Beatley & Manning, *The Ecology of Place: Planning for Environment, Economy, and Community*, 141
Begley, *The Truth About Denial*, 8
Bello, *Cities Cultivate 2 Types of Green*, 61
Benedict XVI, Pope, 204
Berger, *Renewable Energy Sources as a Response to Global Climate Concerns*, 90, 111, 120, 121, 122, 124, 125, 126, 163
Berkeley, California
  Measure G, 54-55, 122, 185
Berlin, 16
Bernow et al., *Carbon Abatement with Economic Growth: A National Strategy*, 18, 120, 124, 125, 126, 189, 193

INDEX 255

Bernstein & Vara-Orta, *Near the Rails but Still on the Road*, 148
Bianchi, *Cross Compliance: The New Frontier in Granting Subsidies to the Agricultural Sector in the European Union*, 32
Biofuel, 24
Biomass, 42
BLACK, INTERNAL COMBUSTION: HOW CORPORATIONS AND GOVERNMENTS ADDICTED THE WORLD AND DERAILED THE ALTERNATIVES, 95, 97, 99, 100, 101, 162, 164, 173, 174, 175, 176, 179
Bluemel, *Unraveling the Global Warming Regime Complex: Competitive Entropy in the Regulation of the Global Public Good*, 23, 32
Böhringer & Finus, *The Kyoto Protocol: Success or Failure?*, 112
Bonta, *The United States Should Not Increase Foreign Aid to Developing Nations*, 66
Boom & Nentjes, *Alternative Design Options for Emissions Trading: A Survey and Assessment of the Literature*, 32
Boston, 17, 120, 144, 145, 169
Boyce, *The Greening of Politics: Seven Years of Rapid Change*, 112
BRAASCH, EARTH UNDER FIRE: HOW GLOBAL WARMING IS CHANGING THE WORLD, 14
Bradsher & Barboza, *Pollution From Chinese Coal Casts a Global Shadow*, 89
BROWN, AMERICAN HEAT: ETHICAL PROBLEMS WITH THE UNITED STATES RESPONSE TO GLOBAL WARMING, 18
Brown, Jerry, 149, 150
Brown, Lester
 Earth Restoration Budget, 57
BROWN, PLAN B 2.0: RESCUING A PLANET UNDER STRESS AND A CIVILIZATION IN TROUBLE, 11, 19, 20, 23, 24, 31, 53, 56, 57, 61, 70, 91, 93, 94, 96, 98, 99, 101, 102, 113, 135, 137, 165, 166, 167, 171, 175, 176, 177, 180, 195, 197, 198, 204
BROWN, GLOBAL WARNING: THE LAST CHANCE FOR CHANGE, 14, 204

Brown, *Ice Caps Melting Fast: Say Goodbye to the Big Apple?*, 16, 17
Brownfields, ch 4, 100-01
 Comprehensive Environmental Response, Compensation, and Liability Act of 1980 (CERCLA), 38
 European brownfield redevelopment, 35-36
 Superfund, 38
BUCHANAN, TEN SHADES OF GREEN: ARCHITECTURE AND THE NATURAL WORLD, 115
Buenos Aires, 17
BUILDING WITHOUT BORDERS: SUSTAINABLE CONSTRUCTION FOR THE GLOBAL VILLAGE, 115
BULKELEY & BEWTSILL, CITIES AND CLIMATE CHANGE: URBAN SUSTAINABILITY AND GLOBAL ENVIRONMENTAL GOVERNANCE, 54, 142, 185
Bureau of Land Management, 111
Burns, Global Climate Change: A Business Perspective, 65
Burtraw & Portney, *A Carbon Tax to Reduce the Deficit*, 103, 114, 124, 125, 189, 190, 192
Busch et al., *Tax and Financial Incentives for Green Building*, 115
Bush, George W., 17, 69, 80, 112, 113
 State of the Union Address, 112
Calgary, Canada, 102
California, 105, 106, 151, 173-74, 195
California Air Resources Board, 149
CALIFORNIA ENERGY COMMISSION, GLOBAL CLIMATE CHANGE POTENTIAL IMPACTS & POLICY RECOMMENDATIONS, 155, 159, 198
Canada, 16, 32, 155
Carbon, ch 2
 Automobiles, 161-62
 Carbon dioxide emissions, 9-10
 Carbon credits, 22-23, 49-50, 114
 Carbon footprint, 43, 46, 93
 Carbon offsets, 49-50
 Carbon tax, 65, 114, 134, 190, 192
 Carbon Trading, 22-23, 32-33, 64, 113-14, 124, 189
 Chicago Climate Exchange, 33

Carbon, *continued*
  Clean Development Mechanism, 192
  Construction, 119-120
  Industry emissions, 41-42
  Leakage, 192
  Personal emissions, 41, 43
  Transportation emissions, 163-64
Carlin, *Global Climate Change Control: Is There a Better Strategy Than Reducing Greenhouse Gas Emissions?*, 5
Carlarne, *Climate Change—The New "Superwhale" in the Room: International Whaling and Climate Change Politics—Too Much in Common?*, 191
Carlson, *Federalism, Preemption, and Greenhouse Gas Emissions*, 187
Carlson & Malloy, *Special Edition: California's AB 1493: Trendsetting or Setting Ourselves Up To Fail?*, 30
Carr & Rosembuj, *Flexible Mechanisms for Climate Change Compliance: Emission Offset Purchases Under the Clean Development Mechanism*, 23
CARSON & VAITHEESWARAN, ZOOM: THE GLOBAL RACE TO FUEL THE CAR OF THE FUTURE, 164, 165, 179, 191
CARSON, SILENT SPRING, 7, 27
Central America, 25, 42
CERVERO, THE TRANSIT METROPOLIS: A GLOBAL INQUIRY, 30, 185
Chaker, *Clothesline Has Neighbors Bent Out of Shape in Bend*, 44
Chamberlain, *Skyfarming*, 20, 21-22
Chang, *Foreign Affairs Federalism: The Legality of California's Link With the European Union Emissions Trading Scheme*, 188-89
Chapman & Davis, *Global Warming—More than Hot Air?*, 9
Charleston, South Carolina, 142
Chemerinsky et al., *California, Climate Change, and the Constitution*, 188
Cheney on Global Warming: Vice President's Views at Odds with Majority of Climate Scientists, 7
Cherry Hill, New Jersey
  Recycling, 55
Chicago, 123, 143, 169
Chicago Climate Exchange, 33

China
  Automobiles, 165
  Birth Control, 139
  Coal-fired power plants, 89
  Energy, 136
  Food, 197
  Gasoline consumption, 165
  Water use, 47
Chloroflurocarbons (CFCs), 9
Chong, *Global Warming: Enough to Make You Sick*, 11
Clean Development Mechanism, 192
CLIMATE CHANGE AND THE KYOTO PROTOCOL: THE ROLE OF INSTITUTIONS AND INSTRUMENTS TO CONTROL GLOBAL CHANGE, 112
CLIMATE CHANGE POLICY: A SURVEY, 15
CLIMATE CHANGE: SCIENCE, STRATEGIES, & SOLUTIONS, 15
CLIMATE CHANGE: WHAT IT MEANS FOR US, OUR CHILDREN, AND OUR GRANDCHILDREN, 8, 15
Clothing, 49
Coal
  Clean coal 91
  Impacts chs2, 7-8, 91
  Indian Coal, 95
  Subsidies, 93-94
  Use, 90-91, 92
The Coal Trap, 96
Cohn, Student Article, *The Brownfields Revitalization and Environmental Restoration Act: Landmark Reform or a "Trap for the Unwary"?*, 38
Colangelo, Comment, *The Politics of Preemption: An Application of Preemption Jurisprudence and Policy to California Assembly Bill 1493*, 188
Commerce Clause, 29, 104, 188
COMMITTEE ON WATERSHED MANAGEMENT, WATER SCIENCE AND TECHNOLOGY BOARD, COMMISSION ON GEOSCIENCES, ENVIRONMENT, AND RESOURCES, NATIONAL RESEARCH COUNCIL, NEW STRATEGIES FOR AMERICA'S WATERSHEDS, 196
Compact fluorescent light bulbs (CFLs), 53-54, 101-102
Congestion pricing, 55

INDEX   257

Congressional Risk Office, 84
Connecticut, 168
Conner, *Managing the Risks of LEED Certification*, 118
Conservation, ch 2, 101-102
   Aerosols, 48, 50, 53
   ATM receipts, 50
   Automatic bank deposits, 50-51
   Compact fluorescent light bulbs (CFLs), 53-54, 101-02
   Burning, 51
   Butane lighters, 51
   Clothing, 49
   Eco-labeling, 56
   Jewelry, 50
   Junk mail, 50
   Lawn fertilizer, 52
   Lights, 48
   On-line bill paying, 50
   Overweight, 49-50
   Telephone books, 51
   See appliances
   See transportation
   See waste
Construction, ch10
   Carbon generation, 119
Copenhagen, Denmark
   Richshaw Bicycle, 177
Cortese, *As the Earth Warms, Will Companies Pay?*, 14
COULTER, GODLESS, 8
Couzin, *Living in the Danger Zone*, 14
Cox, *SUVs Without Wheels*, 119-120
CRAWFORD, CARFREE CITIES, 185
CRICHTON, STATE OF FEAR, 8
Cutraro, *Engineering Change: A Brief History of the Creation and Growth of the Army Corps*, 200
Dallas, Texas, 89
D'Ambrosio, Student Article, *Alternative Fuels: An Evaluation of Corn Ethanol, Cellulosic Ethanol, and Gasoline*, 176
Dark Winter, 79
DASSLER & PARSON, THE SCIENCE AND POLITICS OF GLOBAL CLIMATE CHANGE: A GUIDE TO THE DEBATE, 4, 8, 11
Dausch, Comment, *Analyzing a Municipality's Authority to Enact the Model Ordinance for Wind Energy Facilities in Pennsylvania*, 102
Davison, *Regulation of Emission of Greenhouse Gases and Hazardous Air Pollutants from Motor Vehicles*, 187-88
Deforestation, ch 2, 32
Delahunty, *Federalism Beyond the Water's Edge: State Procurement Sanctions and Foreign Affairs*, 188
Delaware, 168
Del Percio, Student Article, *The Skyscraper, Green Design & LEED Green Building Rating System: The Creation of Uniform Sustainable Standards for the 21st Century or the Perpetuation of an Architectural Fiction?*, 116
Delucchi, Mark, 166-67
Delucchi Study Finds that U.S. Motorists Do Not Pay Their Way, 166-67
DeMeo et al., *Insuring Against Environmental Unknowns*, 38
Dempsey, *Coalition Tackles Germany's Falling Birth Rate: Financial Package for Working Women Wins Wide Support*, 137
Department of Homeland Security, 77-78
DEPARTMENT OF HOMELAND SECURITY, NATIONAL RESPONSE FRAMEWORK, 76
Dernbach, *Overcoming the Behavioral Impetus for Greater U.S. Energy Consumption*, 89
Dernbach, *Sustainable Development as a Framework for National Governance*, 142
Dernbach, *U.S. Policy*, 33, 43, 100
DeShazo & Freeman, *Timing and Form of Federal Regulation: The Case of Climate Change*, 188
Detroit Edison, 60
District of Columbia, 105, 121
Donehower, Comment, *Analyzing Carbon Emissions Trading: A Potential Cost Efficient Mechanism to Reduce Carbon Emissions*, 23, 33, 65, 113, 114, 124, 189
Douglass, *His Passion for Solar Still Burns*, 116
Douma, *The European Union, Russia and the Kyoto Protocol*, 113

Downey, *Smart Growth Policy Revision Needed, Some Say*, 150
Doyle, *L.A.'s Trash Goal: No Waste by 2030*, 61
Driving to Green Buildings: The Transportation Energy Intensity of Buildings, 118, 142, 164
DTE Energy, 60
Earth Restoration Budget, 57
EBERHART, FEEDING THE FIRE: THE LOST HISTORY & UNCERTAIN FUTURE OF MANKIND'S ENERGY ADDICTION, 89, 90
Ecological footprint, 42
Economic Development, ch6
Edmonds et al., *International Emissions Trading*, 64
Education, ch7
  Apprenticeship, 69, 72
  Universities, 69
EHRLICH, THE POPULATION BOMB: POPULATION CONTROL OR RACE TO OBLIVION?, 137, 139
Eisen, *"Brownfields of Dreams"?: Challenges and Limits of Voluntary Cleanup Programs and Incentives*, 35
ELLIOT ET AL, AN ASSESSMENT OF THE AVAILABLE WINDY LAND AREA AND WIND ENERGY POTENTIAL IN THE CONTIGUOUS UNITED STATES, 98
El Niño, 98
Emergency Preparedness, ch 8
  Congressional Risk Office, 84
  Department of Homeland Security, 77-78
  Federal Emergency Management Agency (FEMA), 78-80, 83-84
  Government Accountability Office, 84
  Incident Command System (ICS), 76
  NATIONAL GUARD, 85
  National Incident Management System (NIMS), 76, 87
  National Response Plan, 78
  SPECIAL POPULATIONS, 81
  TRUST FOR AMERICA'S HEALTH, 81
Emerson, *Making Main Street Legal Again: The SmartCode Solution to Sprawl*, 142
Emily, *Chicago Green Roof Program*, 123

Energy,
  Biomass, 100
  Consumption, 89-90
  Cooking, 44-45
  Cost, 43
  Generally, ch 9
  Geothermal Energy, 100
  Green power options, 52-53
  Hydroelectric power, 96
  NUCLEAR POWER, 92, 94, 96-97
  SOLAR POWER, 97, 99-100
  TIDAL TURBINES, 100
  Wind Energy, 92, 97-99
Endangered Species
  Polar Bears, 153-54
Energy Star, 43, 54, 56
Engel, *Harmonizing Regulatory and Litigation Approaches to Climate Change Mitigation: Incorporating Tradable Emissions Offsets Into Common Law Remedies*, 150
Engel, *State and Local Climate Change Initiatives: What is Motivating State and Local Governments to Address a Global Problem and What Does This Say About Federalism and Environmental Law?*, 30, 187
Enrich, *Business Tax Incentives: A Status Report*, 64
Ethanol, 23-24, 100, 112, 175-176
EU CLIMATE CHANGE POLICY: THE CHALLENGE OF NEW REGULATORY INITIATIVES, 15
Europe, 16, 22, 24, 47, 97
  Brownfield redevelopment, 35-36
  Public Transport, 52
European Union, 32, 97
Evans-Pritchard, *Why the Price of 'Peak Oil' is Famine*, 24, 100, 135
EWING ET AL., GROWING COOLER: THE EVIDENCE ON URBAN DEVELOPMENT AND CLIMATE CHANGE, 116-117, 142-44, 146-47, 162, 164
Executive Summary, Surface Temperature Reconstructions for the Last 2,000 Years, 15
Farber, *Adapting To Climate Change: Who Should Pay*, 60

INDEX 259

Farber, *Basic Compensation for Victims of Climate Change*, 18, 150
Farber et al., *Reinventing Flood Control*, 84
Farming, ch 2
   Fertilization, 28
   Irrigation, 28, 196
   Manure digesters, 28
   Pesticides, 25, 27
   Herbicides, 25-26
   Urban vertical farming, 20
Faure, *Insurability of Damage Caused by Climate Change: A Commentary*, 14
Federal Emergency Management Agency (FEMA), 78-80, 83-84
Fed. Emergency Mgmt. Agency, About FEMA: What We Do, 83
FEDERAL LAND MANAGEMENT AGENCIES, 127
Feds Allowing Tarsands to Become 'Most Destructive Project on Earth': Report, 177
Felten, *Brownfield Redevelopment 1995–2005: An Environmental Justice Success Story?*, 35
Ferrey, *Converting Brownfield Environmental Negatives into Energy Positives*, 37, 101
Ferrey, *Nothing But Net: Renewable Energy and the Environment, MidAmerican Legal Fictions, and Supremacy Doctrine*, 105
Ferrey, *Sustainable Energy, Environmental Policy, and States' Rights: Discerning the Energy Future Through the Eye of the Dormant Commerce Clause*, 104
FIELD ET AL., CONFRONTING CLIMATE CHANGE IN CALIFORNIA: ECOLOGICAL IMPACTS ON THE GOLDEN STATE, 155-56, 159, 198
Fire Up the Microwave, 44
Fishing, 132
   Salmon, 154-55
   Shrimp, 155
   Subsidies, 132
Fisk, *The Environmentalist's New Clothes*, 49
FLANNERY, THE WEATHERMAKERS: HOW MAN IS CHANGING THE CLIMATE AND WHAT IT MEANS FOR LIFE ON EARTH, 4, 9-11, 13-14, 42-43, 51, 53, 60, 90-93, 172, 178-180
FLEXIBILITY IN CLIMATE POLICY: MAKING THE KYOTO MECHANISMS WORK, 112
Flooding, ch 2
Florida, 177
Flowind, Altamont Pass, California, 99
Food, ch 2
   Cooking, 44-45
   Fishing, 132
   Hydroponics, 196
   Individual wrapped portions, 45
   Local 28-29
   Meat production, 496
   Prepackaged, 50
   Waste, 45
Footprints
   Carbon, 43, 46
   Ecological, 42, 93
Foreign Aid, 32, 56-57, 65
Foster, *The City as an Ecological Space: Social Capital and Urban Land Use*, 122
France, 190
   Water use, 47
France Slaps Penalties on Gas-Guzzling Cars, 190
Frederick & Gleick, *Potential Impacts on U.S. Water Resources*, 12, 198
Freiburg, Germany
   Tram, 184
FREILICH, FROM SPRAWL TO SMART GROWTH: SUCCESSFUL LEGAL, PLANNING, AND ENVIRONMENTAL SYSTEMS, 141
FRIEDMAN, THE WORLD IN FLAT: A BRIEF HISTORY OF THE TWENTY-FIRST CENTURY, 69-71
Fuels
   Biofuel, 24
   Coal, 90-94
   Ethanol, 23-24, 100-12, 176-76
   Fossil Fuel Subsidies, 171
   Gas-fired power plants, 91
   Hybrid, 51-53, 179-80
   Hydrogen Cells, 165, 177-79
   Nuclear power, 92, 94, 96-97
   Oil, 92, 93-94, 162
   Palm Oil, 176-77
   Waste vegetable oil biofuel, 101

Fuller, *Wind Energy Development on BLM Lands*, 111
Galster et al, *Wrestling Sprawl to the Ground: Defining and Measuring an Elusive Concept*, 141
Galveston, Texas, 79
Garcia, *The United States Should Not Support Family Planning Services in Developing Nations*, 137
Gaston, *Taking the Gloves Off of Homeland Security: Rethinking the Federal Framework for Responding to Domestic Emergencies*, 77
GELBSPAN, BOILING POINT: HOW POLITICIANS, BIG OIL AND COAL, JOURNALISTS, AND ACTIVISTS ARE FUELING THE CLIMATE CRISIS—AND WHAT WE CAN DO TO AVERT DISASTER, 11-12, 14, 94, 132, 153, 180
GELBSPAN, THE HEAT IS ON: THE HIGH STAKES BATTLE OVER EARTH'S THREATENED CLIMATE, 8, 22, 67, 94
General Motors
    Hydrogen research, 179
Gentleman, *Architects Aren't Ready for an Urbanized Planet*, 119
Gertner, *The Future is Drying Up*, 195, 197
Gertz, *Tempting Fate: Fifteen Years After the Great Flood of 1993, Floodplain Development is Booming*, 200
Geysers geothermal power plant, California, 92
Gillis, *Sixth Circuit Bans Ohio Tax Credit Under the Commerce Clause, Casting a Pall on Incentives*, 64
Gilpatrick, *Cities Push Tap Water as 'Better Than Bottled,'* 45, 46
GIPE, WIND POWER: RENEWABLE ENERGY FOR HOME, FARM, AND BUSINESS, 98
Glicksman, *Global Climate Change and the Risks to Coastal Areas from Hurricanes and Rising Sea Levels: The Costs of Doing Nothing*, 11
Global climate change, ch 2
    Costs, ch 2
    Consequences ch 2
    Science, ch 2

GLOBAL DEVELOPMENT OF ORGANIC AGRICULTURE: CHALLENGES AND PROSPECTS, 20
GLOBAL ENVIRONMENTAL OUTLOOK (GEO$_4$), 9, 17, 59, 92
Global Facility for Reduction and Recovery at the World Bank, 87
Global Warming, 7
GLOVER, POSTMODERN CLIMATE CHANGE, 15
Go Green Save Money, 45, 48, 50, 51
Goldberg & Delfino, *The Impact of the Kyoto Protocol on U.S. Business*, 60
Golden Gate Bridge, 51
GOLDSTEIN, SAVING ENERGY GROWING JOBS: HOW ENVIRONMENTAL PROTECTION PROMOTES ECONOMIC GROWTH, PROFITABILITY, INNOVATION, AND COMPETITION, 54, 56, 60, 63, 125
Gooch, *Fenced In: Why Sheff v. O'Neill Can't Save Connecticut's Inner City Schools*, 70
GOODELL, BIG COAL: THE DIRTY SECRET BEHIND AMERICA'S ENERGY FUTURE V, 8, 60, 62, 90, 91, 93, 180
Goodell, *The Prophet of Climate Change: James Lovelock*, 16, 27, 97, 198, 203
Goodell, *The Ethanol Scam: One of America's Biggest Political Boondoggles*, 176
GOODSTEIN, OUT OF GAS: THE END OF THE AGE OF OIL, 96-97, 136, 165, 177
Gordon, *'Green' Builders May Get Fast Track*, 121
Gordon, *Climate Change and the Poorest Nations: Further Reflections on Global Inequality*, 12
Gore, Al, vii, xiii, 4, 49, 103, 161
GORE, AN INCONVENIENT TRUTH: A PLANETARY EMERGENCY OF GLOBAL WARMING AND WHAT WE CAN DO ABOUT IT, 4, 7, 9, 10, 11, 12, 14, 18, 41, 64, 103, 161
Goulder & Nadreau, *International Approaches to Reducing Greenhouse Gas Emissions*, 64-65, 103, 104, 112, 113, 114, 124, 125, 126, 159, 189, 190, 191, 192
Government Accountability Office, 84

Graditor, Comment, *Responsibility for the Restoration of the Hurricane Insurance Industry: Business Proposal or State Solution?*, 13
Green Architecture, ch 10
  Leadership in Energy and Environmental Design ("LEED"), 118
Greenberger, *Preparing Vulnerable Populations for a Disaster: Inner-City Emergency Preparedness—Who Should Take the Lead?*, 76
Greenhouse gases, ch 2
Greenland, 16
  Icecap, 16
Green Site Planning, 116
Greve & Klick, *Preemption in the Rehnquist Court: A Preliminary Empirical Assessment*, 188
Group of Eight Presidencies (G8), 17
Grunwald, *Cry Me a River*, 200
Grunwald, *Setting the Stage for More Katrinas*, 86, 200
Guhathakurta & Patricia Gober, *The Impact of the Phoenix Urban Heat Island on Residential Water Use*, 196
Gulf of Mexico, 24
Gumbel, *Pasta Panic: The Price of Wheat is Up 60% This Year, and in Italy They're Taking to the Streets Over the Cost of Tortellini*, 176
Guzy, *Insurance and Climate Change*, 81
HALL, GREAT PLANNING DISASTERS, 3
HALLETT, THE SOCIAL ECONOMY OF WEST GERMANY, 70
Hanley, *'Drilling Up' Into Space for Energy*, 5
Hansen, *The Threat to the Planet*, 203
Haque, Note, *Internal Revenue Code Section 1989, the Tax Incentive for Brownfield Redevelopment: A Sheep in Wolf's Clothing*, 38
Hardt, *Federal Land Management in the Twenty-First Century: From Wise Use to Wise Stewardship*, 127
Haroff & Katherine Kirwan Moore, *Global Climate Change and the National Environmental Policy Act*, 150
Harper, *Climate Change and Tax Policy*, 8, 124, 125, 189, 190, 192

Harrington, *Presidential Powers Revisited: an Analysis of the Constitutional Powers of the Executive and Legislative Branches over the Reorganization and Conduct of the Executive Branch*, 80
Harris, *Collective Action on Climate Change: The Logic of Regime Failure*, 59
Harris, *Lawn to Farm: Suburbia's Silver Lining*, 27
Hawaii, 110
HAWKES & FORSTER, ENERGY EFFICIENT BUILDINGS: ARCHITECTURE, ENGINEERING, AND ENVIRONMENT, 115
Hayes & Beauvais, *Carbon Sequestration*, 9
HEARNDEN, EDUCATION, CULTURE, AND POLITICS IN WEST GERMANY, 70
HEINBERG, POWER DOWN: OPTIONS AND ACTIONS FOR A POST-CARBON WORLD, 166, 179
Heinmiller, *The Politics of "Cap and Trade" Policies*, 23
Heinzerling, *Climate Change and the Clean Air Act*, 8
Heinzerling, *Climate Change, Human Health, and the Post-Cautionary Principle*, 86
Heinzerling & Ackerman, *Law and Economics for a Warming World*, 3, 14, 75, 132, 151
Heliostats at the Solar Two Power Plant, Daggett, California, 98
HENSON, THE ROUGH GUIDE TO CLIMATE CHANGE: THE SYMPTOMS, THE SCIENCE, THE SOLUTIONS, 10, 15, 113
Hersch & Viscusi, *Allocating Responsibility for the Failure of Global Warming Policies*, 150
Hill et al., *Environmental, Economic, and Energetic Costs and Benefits of Biodiesel and Ethanol Biofuels*, 100
HILLMAN ET AL, THE SUICIDAL PLANET: HOW TO PREVENT GLOBAL CLIMATE CATASTROPHE, 20, 89, 90, 92, 96, 161, 164, 170, 172
HIRO, BLOOD OF THE EARTH: THE BATTLE FOR THE WORLD'S VANISHING OIL RESOURCES, 93, 95, 162, 165, 179, 180

Hirsch et al., *Emissions Trading—Practical Aspects*, 23
Hodas, *State Initiatives*, in GLOBAL CLIMATE CHANGE AND U.S. LAW, 106, 107, 110
Hoffman & Coler, *Brownfields and the California Department of Toxic Substances Control: Key Programs and Challenges*, 35
Holtkamp, *Dealing with Climate Change in the United States: The Non-Federal Response*, 104
Honda
  Hydrogen research, 179
Hosey, *Is Bigger Better?: New Research Supports Eco-Friendly Urbanism*, 143
HOUGHTON, GLOBAL WARMING: THE COMPLETE BRIEFING, 9, 10, 11, 12, 15, 32, 104
Houston, 144
Hunter & Salzman, *Negligence in the Air: The Duty of Care in Climate Change Litigation*, 150-51
Hwang, *The United States Should Support Family Planning Services in Developing Nations*, 137
Housing, ch 10
  Carbon generation, 119-20
Hsiung & Sunstein, *Climate Change and Animals*, 153
Hughes & Taylor, *And Can't Look Up and See the Stars*, 48
Hunter & Salzman, *Negligence in the Air: The Duty of Care in Climate Change Litigation*, 150-51
Hwang, *The United States Should Support Family Planning Services in Developing Nations*, 137
Hydrogen cells, 165, 177-78
Hydrologic cycle, ch 2, 195-96
Icecaps, 11, 16
Iceland, 16
Idaho, 106, 109-110
Illinois, 30, 106
  Response to Governor's Sustainable Energy Plan for the State of Illinois, 105
Ilulissat Ice Fjord Glacier, 16

In Hot Seat, Bush Unveils New Climate Strategy, 17
Incident Command System (ICS), 76
India, 136
  Automobiles, 165
INNOVATION IN ENERGY TECHNOLOGY: COMPARING NATIONAL INNOVATION SYSTEMS AT THE SECTORAL LEVEL, 93
Insurance, 13-14, 81
Inter-Agency Secretariat of the International Strategy for Disaster Reduction, 87
Intergovernmental Panel on Climate Change (IPCC), 3-4, 16
INTERGOVERNMENTAL PANEL ON CLIMATE CHANGE, CLIMATE CHANGE 2007, THE PHYSICAL SCIENCE BASIS, SUMMARY FOR POLICY MAKERS, 15
INTERGOVERNMENTAL PANEL ON CLIMATE CHANGE, CLIMATE CHANGE 2001, 8
INTERNATIONAL CITY/COUNTY MANAGEMENT ASS'N, MEASURING SUCCESS IN BROWNFIELDS REDEVELOPMENT PROGRAMS, 35
Iowa, 44
Iran, 69
Iraq, 75, 167
Irrigation, 24
  Drip irrigation, 47
Jackson, *Farmer Take Another Look at Wind Energy*, 19, 98
JACQUES, GLOBALIZATION AND THE WORLD OCEAN, 132
Jaffe, *Global Warming?*, 7
Jakarta, 17
Jamieson, *Ethics, Public Policy and Global Warming*, 18
Japan, 44
JOHANSEN, GLOBAL WARMING IN THE 21ST CENTURY, 3, 8, 10, 11, 13, 15, 18, 22, 27, 113, 153, 171, 172, 178
Johnson, *Shining a Light on Hazards of Fluorescent Bulbs: Energy-Efficient Coils Booming, But Disposal of Mercury Poses Problems*, 53
Johnston, *Desperately Seeking Numbers: Global Warming, Species Loss, and the Use and Abuse of Quantification in Climate Change Policy Analysis*, 153

Johnstone, *Tradable Permits for Climate Change: Implications for Compliance, Monitoring, and Enforcement*, 23, 32
Jones, *It Won't Be Easy Being Green: Berkeley Sets Tough Course for its Residents to Follow to Help Reduce Emissions of Greenhouse Gases in City*, 54, 55, 122, 185
Jonsson, *Planners Raise Local Funds for Innovative Projects Instead of Relying on State and Federal Money*, 174
Jordanian desert, 97
Jorissen, Note, *Katrina's House: The Constitutionality of the Forced Removal of Citizens from their Homes in the Wake of Natural Disasters*, 81-82
KAHN, GREEN CITIES: URBAN GROWTH AND THE ENVIRONMENT, 10, 18, 143, 144, 145
Kaiser, *Money—With Strings—to Fight Poverty*, 66, 73
Kamarck, *When First Responders Are Victims: Rethinking Emergency Response*, 77, 78, 79
Kamenetz, *The Green Standard?: LEED Buiuldings Get Lots of Buzz, But the Point is Getting Lost*, 118
Karachi, 17
Kaswan, *The Domestic Response to Global Climate Change: What Role for Federal, State, and Litigation Initiatives?*, 103, 111
Katrina, Hurricane, 9, 76, 77, 78, 79
Keeana, Comment, *The Terminator a Trendsetter? How California's Global Warming Solutions Act Will Impact California, the United States, and the World*, 149
KELBAUGH, COMMON PLACE: TOWARD NEIGHBORHOOD AND REGIONAL DESIGN, 3
Kerr et al., *Latest Forecast: Stand By for a Warmer But not Scourching, World*, 18
Key West, Florida, 142
Kibert & Grosskopf, *Envisioning Next-Generation Green Buildings*, 116
King & King, *Creating Incentives for Sustainable Buildings: A Comparative Law Approach Featuring the United States and the European Union*, 118, 122

Kingsley, *Making it Easy to be Green: Using Impact Fees to Encourage Green Building*, 115-16, 123
Klein, *The New Nuisance: An Antidote to Wetland Loss, Sprawl, and Global Warming*, 151
Kluger, *Global Warming Heats Up*, 8
Knaap, & Frece, *Smart Growth in Maryland: Looking Forward and Looking Back*, 142
KNOX & FOLEY, GLOBAL CLIMATE CHANGE AND CALIFORNIA: POTENTIAL IMPACTS AND RESPONSES, 156, 159, 198
Kojo, *The Importance of the Geographic Origin of Agricultural Products: A Comparison of Japanese and American Approaches*, 28
Kolkata, 17
Kosloff & Trexler, *Consideration of Climate Change in Facility Permitting*, 102
KOSLOW, THE SILENT DEEP: THE DISCOVERY, ECOLOGY AND CONSERVATION OF THE DEEP SEA, 131, 132, 134
Kunreuther & Michel-Kerjan, *Climate Change, Insurability of Large-Scale Disasters, and the Emerging Liability Challenge*, 14
KUNSTLER, THE LONG EMERGENCY: SURVIVING THE CONVERGING CATASTROPHES OF THE TWENTY-FIRST CENTURY, 20, 69, 90, 93, 97, 100, 161, 164, 165, 178
Kurdila & Rindfleisch, *Funding Opportunities for Brownfield Development*, 36
Kushner, *Brownfield Redevelopment Strategies in the United States*, 35, 36
Kushner, *Car Free Housing Developments: Towards Sustainable Smart Growth and Urban Regeneration Through Car-Free Zoning, Car-Free Redevelopment, Pedestrian Improvement Districts, and New Urbanism*, 183, 185
Kushner, *Growth for the Twenty-First Century: Tales from Bavaria and the Vienna Woods—Comparative Images of Urban Planning in Munich, Salzburg, Vienna, and the United States*, 70
KUSHNER, HEALTHY CITIES—THE INTERSECTION OF URBAN PLANNING, LAW, AND HEALTH, 143, 183

Kushner, The Post-Automobile City, 30, 54, 122, 143, 183, 185, 186
Kushner, *Smart Growth, New Urbanism, and Diversity: Progressive Planning Movements in America and Their Impact on Poor and Minority Ethnic Populations*, 142
Kushner, *Social Sustainability: Planning for Growth in Distressed Places—the German Experience in Berlin, Wittenberg, and the Ruhr*, 36
Kushner, Subdivision Law and Growth Management, 30
Kyoto Protocol, 18, 22, 60, 112-14, 180, 191-93
Lagos, 17
Lake Benton, Minnesota, 158
Lamm, *Immigration: The Ultimate Environmental Issue*, 86
Lancaster, Transforming Foreign Aid: United States Assistance in the 21st Century, 66
Lang & Nelson, *America 2040: The Rise of the Megapolitans*, 135, 161, 182
Latin America, 135-36
The Law of the Sea: Progress and Prospects, 132
Layton & Hsu, *Letting the Market Drive Transportation: Bush Officials Criticized for Privatization*, 162
Leadership in Energy and Environmental Design ("LEED"), 117-18
Lederer, *Biofuel Growth Adds to Hunger: Most Vulnerable 'Priced Out' of Market for Food*, 23, 176
Lederer, *UN Expert Decries Turning Food Into Fuel*, 23
A Legal Guide to Homeland Security and Emergency Management for State and Local Governments, 76
Leonard, *A Plague of Bloodsuckers: In Japan, Reforestation, Population Decline and Global Warming Have Set Off a Land Leech Invasion*, 22
Leroux, Global Warming—Myth or Reality? The Erring Ways of Climatology, 8
LeRoy, *Compulsory Labor in a National Emergency: Public Service or Involuntary Servitude? The Case of Crippled Ports*, 81
Lewyn, *How Government Regulation Forces Americans into their Cars: A Case Study*, 166
Lewyn, *Sprawl, Growth Boundaries and the Rehnquist Court*, 37, 141
Lewyn & Cralle, *Planners Gone Wild: The Overregulation of Parking*, 166, 169
Lib Dems Plan Air Tax to aid rail, 191
Lima, 17
Linn, *Carbon Offset Market Raises Questions*, 49
Liu, *Education, Equality, and National Citizenship*, 70
Livi-Bacci: A Concise History of World Population, 135
Local food, 29
Loma Prieta earthquake, 168
London, 36
   Heathrow Airport, 185-86
   Thames Barrier, 82
   Ultra Light Transport, 185-86
Long Beach, California, 148
   Blue Line Light Rail, 148
Longhurst, Ecological Geography of the Sea, 132
Los Angeles, 143
   Blue Line Light Rail, 148
   Parking, 169
   Traffic, 162
Loucks, *Business Capitalizing on Energy Transition Opportunities*, 14, 114, 125
Lovelock, Gaia: A New3 Look at Life on Earth, 16
Lovelock, Healing Gaia: Practical Medicine for the Planet, 16
Lovelock, James, 16, 203
Lovins, *Winning the Oil Endgame: Innovation for Profits, Jobs, and Security*, 164
Lowe, *Cars, Their Problems, and the Future*, 164, 168, 182
MacDonald & Makuch, *Emissions Trading and the Aarhus Convention: A Proportionate Symbiosis?*, 65, 113
Macht, *The Rise of Car Sharing*, 54, 122, 185
Maine, 111

Management of Federal Lands and Agencies, ch 11
Manale, *Agriculture and the Developing World: Intensive Animal Production, a Growing Environmental Problem*, 21
Mandelbaum, *Corporate Sustainability Strategies*, 118
Mander, *Globalization is Harmful to the Environment*, 172
Manhattan, 169
Manhattan Parking Spot Going for $225,000, 169
Manila, 17
Mank, *Reforming State Brownfield Programs to Comply with Title VI*, 38
Mann, *Another Day Older and Deeper in Debt: How Tax Incentives Encourage Burning Coal and the Consequences for Global Warming*, 90, 103
Mann, *Subsidies, Tax Policy, and Technological Innovation*, 95
Manning, Against the Grain: How Agriculture Has Hijacked Civilization, 19
Mannion, *In it for the Short Haul: Car-Sharing for Urban Errands Brings it to Battle Against Price, Parking and Pollution to Chicago*, 54-55, 122, 185
Marble Road, Ephesis, Turkey, 205
Margolis & Kammen, *Energy R&D and Innovation: Challenges and Opportunities*, 122, 126
Marr, Precautionary Principle in the Law of the Sea: Modern Decision-making in International Law, 132
Martin, The Meaning of the 21st Century, 11, 12, 14, 19, 21, 24, 25, 26, 32, 42, 47, 70, 89, 94, 132, 137, 139, 155, 195, 196, 197
Martin, *Sprawl Clashes with Warming in California*, 149
Marty, *Hurricane Katrina: A Deadly Warning Mandating Improvement to the National Response to Disasters*, 75-76
Mason, Planet Under Pressure—Population, 46, 135
Massachusetts, 145, 151
Mazzetti & Havemann, *Iraq War is Costing $100,000 per Minute*, 167

Maugh & Kaplan, *Katrina Leaves Permanent Scar on Forests*, 9
McConnell, *Annual Threat Assessment of the Director of National Intelligence*, 176
McGavin, Endangered: Wildlife on the Brink of Extinction, 153
McKibben, Hope, Human and Wild: True Stories of Living Lightly on the Earth, 63, 183
McKie, *How Africa's Desert Sun Can Bring Europe Power*, 97
McKinstry, Jr., *Laboratories for Local Solutions for Global Problems: State, Local and Private Leadership in Developing Strategies to Mitigate the Causes and Effects of Climate Change*, 104
McMorrow, Note, *CERCLA Liability Redefined: An Analysis of the Small Business Liability Relief and Brownfields Revitalization Act and its Impact on State Voluntary Cleanup Programs*, 38
McNulty, *Green Leaves, Black Gold*, 177
Mexico, 66, 73
Meyerson, *Population and Climate Change Policy*, 9, 12, 136, 137
Miami, 17
Miami Beach, 17
Middle East, 167
Millennium Housing, London, 36
Miller, *Climate Change and Water in the West: Complexities, Uncertainties and Strategies for Adaptation*, 11
Miller, *International Trade and Development*, 66
Miskel, *Disaster Response and Homeland Security: What Works, What Doesn't*, 76, 83, 85, 86
Mississippi, 83
Mohl, *More Insurers Backing Away From Coasts*, 14
Monbiot, Heat: How to Stop the Planet From Burning, 8, 15, 17, 18, 89, 91, 112, 113, 116, 157, 164, 167, 172, 180, 181, 191, 203
Moran, Harnessing Foreign Direct Investment for Development: Policies for Developed and Developing Countries, 66

Morgan, Note, *Carbon Trading Under the Kyoto Protocol: Risks and Opportunities for Investors*, 33
MORRIS, ENERGY SWITCH: PROVEN SOLUTIONS FOR A RENEWABLE FUTURE, 96-97, 99, 100, 105, 115, 178
Morris, *Warming Aid, Chilling Trade?*, 114
Mosquitoes, 12
Mt. St. Helens, 9
Msangi & Rosegrant, *Agriculture and the Environment: Linkages, Trade-offs and Opportunities*, 23, 175
MSNBC, *U.N. Issues Landmark Report on Global Warming: Panel Offers Dire Warnings, Establishes Scientific Baseline for Political Talks*, 4
THE MULTI-GOVERNANCE OF WATER: FOUR CASE STUDIES, 196
Mumbai, 17
Munich Reinsurance, 14
Nasser, *Senior Transportation a Growing Concern*, 147
National Association of Realtors, 146, 189
National Ass'n of Realtors, News Release, *Americans Prefer to Spend More on Mass Transit and Highway Maintenance, Less on New Roads*, 146, 189
NATO, 64
National Guard, 85
National Incident Management System (NIMS), 76, 87
National Response Plan, 78
National Surface Transportation Policy and Revenue Commission, 167
Navarro, Comment, *What About the Polar Bears? The Future of the Polar Bears as Predicted by a Survey of Success under the Endangered Species Act*, 154
Neibauer, *Legislation to Ease Parking Crunch—for Bicycles*, 186
NELSON, DAWKINS & SANCHEZ, THE SOCIAL IMPACTS OF URBAN CONTAINMENT, 120
Netherlands, 170
Neuman, *Mixed Signals: Driving to Work as a Tax Break*, 170, 171
Nevada, 110
New Jersey, 167

New Jersey Town Doubles Recyling Rates in One Week with the RecycleBank Program, 55
New Orleans, 17, 77, 86, 204
Levee failure, 86
Newtok, Alaska, 18
New Virus Extends Geographic Reach, 12
New York, 16, 17, 27, 73, 143, 167, 177
Greenhouse gas emissions, 143
Parking, 169
Niles, *Tropical Forests and Climate Change*, 12, 31, 32
Nolan, *Disaster Mitigation Through Land Use Strategies*, 78, 82, 84
Nordberg, Note, *Excuse Me, Sir, But Your Climate's on Fire: California's S.B. 1368 and the Dormant Commerce Clause*, 29, 104, 149
Norquist, *We Would Use Less Energy Living Closer Together: Cities Have Powerful Environmental Advantages: They Make it Easier to Walk and Use Public Transit*, 143
North America, 24, 44-45
North Shore, Oahu, Hawaii, 133
Note, *The Compact Clause and the Regional Greenhouse Gas Initiative*, 105
Note, *Making Mixed-Income Communities Possible: Tax Base and Class Desegregation*, 30, 37, 63, 72, 148
Obesity, 49-50
Oceans and Seas, ch 12
Ohshita, *The Scientific and International Context for Climate Change Initiatives*, 15
Oil, 92
Consumption, 162, 164
Cost, 162, 171
Peak Oil, 165-66
Social costs, 173-74
Oliver, All About: Cities and Energy Consumption, 89, 91, 92, 117, 135
Olson, *Fare-Free Public Transit Could be Headed to a City Near You*, 183
Openchowski, *The Next Greenhouse Gas Executive Order?*, 111
Opper, *The Brownfield Manifesto*, 35
Oregon, 110, 111

Oreskes, *Beyond the Ivory Tower: The Scientific Consensus on Climate Change*, 8
Oreskes, *The Scientific Consensus on Climate Change: How Do We Know We're Not Wrong?*, 8
ORFIELD, AMERICAN METROPOLIS: THE NEW SUBURBAN REALITY, 30, 37, 63, 72, 148
Osofsky, *Local Approaches to Transnational Corporate Responsibility: Mapping the Role of Subnational Climate Change Litigation*, 150
Pace Law School Center for Environmental Legal Studies, *State Response to Climate Change: 50-State Survey*, 107, 108, 110
PAHL, THE CITIZEN-POWERED ENERGY HANDBOOK: COMMUNITY SOLUTIONS TO A GLOBAL CRISIS, 20, 97, 98, 100, 101, 165
Pallemaerts & Williams, *Climate Change: The International and European Policy Framework*, 112
Parking, 55, 168-69
  Subsidies, 169, 171
Parry, *Fiscal Interactions and the Case for Carbon Taxes Over Grandfathered Carbon Permits*, 103, 114
Parry & Safirova, *Pay as You Slow: Road Pricing to Reduce Traffic Congestion*, 30, 185
Pasadena, California, 186
Peak Oil, 165-66
Pearson, *Climate Change and Agriculture: Mitigation Options and Potential*, 28
Peeters, *Enforcement of the EU Greenhouse Gas Emissions Trading Scheme*, 65, 113
Peirce, *Follow British Model on Transportation Needs: It's Time for a Big New Tax in America*, 187
Peirce, *Study Shows High Sea Rise Danger for U.S. Coastal Cities*, 17
Percy, *Delta Levees—Tort Immunity vs. Takings Liability*, 83
Peters, Mary, 168, 170
Peters, *Gas Taxes Are High Enough*, 168

PEW CENTER ON GLOBAL CLIMATE CHANGE, LEARNING FROM STATE ACTION ON CLIMATE CHANGE, 125
Phoenix, Arizona, 144
Pierre & Stephenson, *After Katrina: A Critical Look at FEMA's Failure to Provide Housing for Victims of Natural Disasters*, 78
PITTOCK, CLIMATE CHANGE: TURNING UP THE HEAT, 11, 15, 21, 104, 166
PLATT, LAND USE AND SOCIETY: GEOGRAPHY, LAW AND PUBLIC POLICY, 127
Ploetz, Note, *Light Pollution in the United States: An Overview of the Inadequacies of the Common Law and State and Local Regulation*, 48
THE POISONED WELL: NEW STRATEGIES FOR GROUNDWATER PROTECTION, 196
Polar Bears, 154
Poliakoff, *Disaster Planning and Recovery*, 76
Population,
  Forecast, 135-36
  Growth, 42, 135
  Generally, 42, ch 13
Portland, Oregon, 142
Posner, *Climate Change and International Human Rights Litigation: A Critical Appraisal*, 151
POSTEL, LAST OASIS: FACING WATER SCARCITY, 196
Poverty, 135
Preer, *Housing Deal Gets Popular: Towns Just Hope State Ponies Up Aid*, 151
PRESCOTT, FEDERAL LAND MANAGEMENT: CURRENT ISSUES AND BACKGROUND, 127
Prindle, *How Energy Efficiency Can Turn 1300 New Power Plants Into 170*, fact sheet, 101
Pring, *A Decade of Emissions Trading in the USA: Experiences and Observations for the EU*, 23, 65, 113
Privileges and Immunities Clause, 29
Public Lands: Use and Misuse, 127
Pullella, *Save the Planet Before it's Too Late, Pope Urges*, 204
PYLE, RAISING LESS CORN, MORE HELL: THE CASE FOR THE INDEPENDENT

268 INDEX

Farm and Against Industrial Food, 26, 32
Quebec to Collect Nation's 1st Carbon Tax: Energy Companies will Pass Cost to Consumers, Say Analysts, 103-04
Rabe, PEW Ctr. On Global Climate Change, Race to the Top: The Expanding Role of U.S. State Renewable Portfolio Standards, 105
Rabe, Statehouse and Greenhouse: The Emerging Politics of American Climate Change Policy, 105
Radvanovsky, Critical Infrastructure: Homeland Security and Emergency Preparedness, 75, 76, 83
Real Climate, 18
Recycling, 37, 46
  Cherry Hill, New Jersey, 55
Redlener, Americans at Risk: Why We Are Not Prepared for Megadisasters and What We Can Do Now, 75, 76, 81, 84, 85, 86
Reforestation, 22
Register, Ecocities: Rebuilding Cities in Balance With Nature, 162, 168, 185
Reisch & Bearden, Superfund and the Brownfields Issue, 35
Reisner, Cadillac Desert: The American West and its Disappearing Water, 196
Rein, *Md. House Approves Cut in Car Pollution*, 187
Renewable Energy: Power for a Sustainable Future, 94
Renewable Resources and Renewable Energy: A Global Challenge, 94
Rethinking *the Kyoto Protocol: Are There Legal Solutions to Global Warming and Climate Change?*, 32-33
Reverse Osmosis and Toray Membrane: Can Desalination Finally Solve Water Scarcity?, 198
Revkin, *Budgets Falling in Race to Fight Global Warming*, 12
Revkin, *Climate Change as News: Challenges in Communicating Environmental Science*, 7
Rhode Island, 168

Ruhl, *Climate Change and the Endangered Species Act: Building Bridges to the No-Analog Future*, 153
Richards, *VTA Finds Hydrogen Buses Cost Much More to Run Than Diesel Vehicles*, 178
Riker, *The Green Zone: Green Building Requirements Must Strike a Balance Between Market Economics and Social Needs*, 118
Rischard, High Noon: Twenty Global Problems, Twenty Years to Solve Them, 15
Rising Sea Levels Send Ripples Through Real Estate Industry, 80, 81
Rita, Hurricane, 78, 79
  Galvaston evacuation, 79
Ritter, *Calif. Sees Sprawl as Warming Culprit*, 149
Robinson, *Fulfilling Brown's Legacy: Bearing the Costs of Realizing Equality*, 70, 72
Rogers & Kostigen, The Green Book: The Everyday Guide to Saving the Planet One Simple Step at a Time, 43-53, 97, 116-18, 170-72, 197
Romm, Hell and High Water: Global Warming—The Solution and the Politics—and What We Should Do, 8, 9, 11, 15, 16, 178, 179
Romm, The Hype About Hydrogen: Fact and Fiction in the Race to Save the Climate, 14, 179, 180
Rosenthal, *U.N. Report Describes Risks of Inaction on Climate Change*, 4
Rosenwald, *The Rising Tide of Corn*, 175
Rusk, Inside Game—Outside Game: Winning Strategies for Saving Urban America, 30, 37, 63, 72, 148
Russia, 69
Russian Region to Host Day of Conception, 137
Russell, *Can LEED Survive the Carbon-Neutral Era?*, 118
Ryan, *Schools, Race, and Money*, 70, 72
Ryan & Heise, *The Political Economy of School Choice*, 70
Sacramento, California, 123, 125
Sahara Desert, 16

INDEX

Sale of Carbon Credits Helping Land-Rich, But Cash-Poor, Tribes, 22
Salkin, *Sustainability at the Edge: The Opportunity and Responsibility of Local Governments to Most Effectively Plan for Natural Disaster Mitigation,* 84
Salsich, Jr., *Toward a Policy of Heterogeneity: Overcoming a Long History of Socioeconomic Segregation in Housing,* 126
Sample, *Global Food Crisis Looms as Climate Change and Population Growth Strip Fertile Land,* 25, 26, 135
San Bernardino County, CALIFORNIA, 149
San Diego, California, 80, 150
San Francisco, 17, 46, 120, 143, 144
San Mateo County, California, 121
Saudi Arabia, 69, 177
Savannah, Georgia 17, 142
SB 375 Connects Land Use and AB 32 Implementation, 151
Scandinavia, 16
SCHABBEL, THE VALUE CHAIN OF FOREIGN AID: DEVELOPMENT, POVERTY REDUCTION, AND REGIONAL CONDITIONS, 65, 66
Schelling, *Economic Responses to Global Warming: Prospects for Cooperative Approaches,* 59
Schneider & Kuntz-Duriseti, *Uncertainty and Climate Change Policy,* 159
Scotland, 154
Schuitema, Comment, *Road Pricing as a Solution to the Harms of Traffic Congestion,* 167
Schwab & Brower, *Increasing Resilience to Natural Hazards: Obstacles and Opportunities for Local Governments Under the Disaster Mitigation Act of 2000,* 84
Schwartz, Note, *Doubtful Duty: Physicians' Legal Obligation to Treat During an Epidemic,* 81
Schwarzenegger, Arnold, 125
Schwinn, 180
Searise
  Projections, 15-17
Seattle, Washington, 143
Secretary Peters Says Bikes "Are Not Transportation," 170
Select Bipartisan Comm. to Investigate the Preparation for and Response to Hurricane Katrina, A Failure of Initiative: Final Report, 78
Selna, *Eco-Tough S.F. Code Proposed for Buildings,* 120, 121
Severn Estuary, 96
Shanghai, 17
Shepherd, *The Future of Technology Transfer Under Multilateral Agreements,* 157
Sheppard, *The Urban Revival: Cities May be the Key to Curbing Climate Crisis,* 143
Shevory, *Homespun Electricity, From the Wind,* 99
SHOUP, THE HIGH COST OF FREE PARKING, 30, 55, 166, 168, 169
Shrimp, 155
Siceloff, *Drivers Might Pay Road Taxes by Mile,* 167
Silicon Valley, 59
SINGER & MASON, THE WAY WE EAT: WHY OUR CHOICES MATTER, 20, 27, 154, 155
SINGER & AVERY, UNSTOPPABLE GLOBAL WARMING: EVERY 1,500 YEARS, 8
Smart Car, Germany, 173
Smart growth
  Generally, ch 14
  Index, 144
  Portland, Oregon, 142
  Subsidies for sprawl, 166
  Urban growth boundaries, 37
Smart Growth America, 146, 189
SMITH & TIRPAK, THE POTENTIAL EFFECTS OF GLOBAL CLIMATE CHANGE ON THE UNITED STATES, 156, 159, 198
SMITH, ARCHITECTURE IN A CLIMATE OF CHANGE: A GUIDE FOR SUSTAINABLE DESIGN, 22, 113, 115, 135
Soares, *Critical Issues in Implementing Energy Taxation,* 103, 114, 124, 189, 190, 192
Solar Power, 97
Solar Car, 158
SOROS ON GLOBALIZATION, 65, 66
Sorrell & Sijm, *Carbon Trading in the Policy Mix,* 23, 32

South America, 42
Spalding & de Fontaubert, *Conflict Resolution for Addressing Climate Change With Ocean-Altering Projects*, 94, 132
Species Protection, ch. 15
Spencer, Note, *Evaluating Kentucky's Investment Tax Credits in Light of Cuno v. DaimlerChrysler, Inc.*, 64
SPETH, RED SKY AT MORNING: AMERICA AND THE CRISIS OF THE GLOBAL ENVIRONMENT, 90, 96, 101, 180
Stacey, *Urban Growth Boundaries: Saying "Yes" to Strengthening Communities*, 37, 141
STEELE, ECOLOGICAL ARCHITECTURE: A CRITICAL HISTORY, 115
Steele-Saccio, *Education by Design*, 118
Stein, *Regional Initiatives to Reduce Greenhouse Gas Emissions*, 103
STERN, STERN REVIEW: THE ECONOMICS OF CLIMATE CHANGE, 15
STIGLITZ & CHARLTON, FAIR TRADE FOR ALL: HOW TRADE CAN PROMOTE DEVELOPMENT, 66
STRATEGIC PLANNING OF SUSTAINABLE URBAN WATER MANAGEMENT, 196
Superdome, New Orleans, 77
Susman & Doll, *Wind Advisory: Finding a Suitable Site for a Wind Farm Requires More Than Locating a Blustery Location*, 98, 102, 110
Sussman, *Reshaping Municipal and County Laws to Foster Green Building, Energy Efficiency, and Renewable Energy*, 102, 120
TAMMINEN, LIVES PER GALLON: THE TRUE COST OF OUR OIL ADDICTION, 26, 41, 52, 94, 161, 162, 163, 165, 167, 173
Tarlock & Van De Wetering, *Water and Western Growth*, 199, 200
Taxes
    Carbon, 65, 114, 134, 190, 192
    Gas Guzzler, 56
    Large homes, 56
Tax base sharing, 30, 37, 63, 72, 148
Tax credits
    Conservation, 44, 124

Technology, ch 16
Technology: Don't Call it a Blimp, 173
TERTZAKIAN, A THOUSAND BARRELS A SECOND: THE COMING OIL BREAK POINT AND THE CHALLENGES FACING AN ENERGY DEPENDENT WORLD, 20, 102, 165, 180, 181
Texas Transportation Institute, 174
Thames Barrier, London, 82
Three Gorges Dam, 96
Time to Tax Carbon, 189, 190, 192
Tolan, *Tax and Insurance Consequences of Major Disasters: Weathering the Storm*, 13
Top 10 Things You Can Do to Reduce Global Warming, 4
Top Tips to Stop Global Warming, 43-44, 46-47, 50-51, 53
Track I, Annex C, International Strategy for Disaster Reduction, 87
Track II, Annex D, International Strategy for Disaster Reduction, 87
Traffic congestion, 165
    Costs, 174
    Congestion pricing, 167-68
    Parking, 55
Tran, *Updated Hurricane Models: A New Opportunity to Insure Against Climate Change*, 14
Transportation
    Air, 49-50, 172
    Automobiles, 49, 51, 161
    Bicycles, 164, 170
    Carbon emissions, 164
    Carpooling, 164
    Commuting, 161
    Congestion pricing, 167-68
    Demand-Response Taxis, 164
    Energy consumption, 163-34
    Generally, ch 17
    Hybrid automobiles, 51, 52-53, 167, 179-80
    Hydrogen cells, 165, 177-78
    Parking, 55, 168-69, 171
    Public transport, 52, 144-45, 181-82
    Rail Freight, 164
    Road Freight, 164
    Sea Shipping, 172

INDEX 271

Subsidies, 166-77
Telecommuting, 52
Train, 164
Travel reduction and conservation, 49-50
Vanpooling, 164
Vans, 164
Walking, 170
Transportation Finance at the Ballot Box: Voters Support Increased Investment & Choice, 145
Trust for America's Health, 81
Tsosie, *Indigenous People and Environmental Justice: The Impact of Climate Change*, 12
Tübingen, Germany, 146
Turman, *Regional Impact Assessments: A Case Study of California*, 12, 155, 159, 198
Turner, *The Secret Greening of Calgary*, 102
"Twenty in Ten" plan, 112
Twenty in Ten: Strengthening America's Energy Security, 112
Union City, Oklahoma Tornado, 13
United Kingdom, 190-91
Water use, 47
United Nations, 17, 23, 64
UNITED NATIONS ENVIRONMENT PROGRAMME, GEO4—GLOBAL ENVIRONMENT OUTLOOK, 9, 17, 59, 92
United States, 32, 155
U.S. Constitution
Commerce Clause, 29, 104, 188
Privileges and Immunities Clause, 29
U.S. Department of Agriculture, *How Anaerobic Digestion (Methane Recovery) Works*, 28
U.S. DEP'T OF ENERGY, ENERGY INFO. ADMIN., ANNUAL ENERGY OUTLOOK, 2006 WITH PROJECTIONS TO 2030, 89
U.S. ENVIRONMENTAL PROTECTION AGENCY, ENERGY STAR, PRODUCT SPECIFICATIONS, ELIGIBILITY CRITERIA & PARTNER COMMITMENTS, 56
U.S. Environmental Protection Agency, Global Warming, Actions, State, 30

U.S. Loans for Coal Plants Clash with Carbon Cuts, 112
U.C. Davis Institute for Transportation Studies, 166-67
Urban growth boundaries, 37
Urban high rise farms, 20
University of Minnesota Solar car, Lake Benton, Minnesota, 158
VALE & VALE, GREEN ARCHITECTURE: DESIGN FOR AN ENERGY-CONSCIOUS FUTURE, 97, 102, 115
Vandenbergh, & Steinemann, *The Carbon-Neutral Individual*, 41-42, 43, 46
Vehicle Miles Traveled, 161-63
Vellinga, *Aspiring to Be America's Greenest City: State Commitment Helps Sacramento Rank No. 2 in Energy Efficient Office Space*, 123, 125
Venezuela, 69
Venus, 10
Vermont, 145
Verticle Farming, 20
Vienna, Austria
Bike Sharing, 175
Vincent et al., Federal Land Management Agencies: Background on Land and Resources Management, 127
Vogel, *Upending the Traditional Farm: Cities are Taking Over Farmland. Could They Someday Take Over the Job of Farming, Too*, 20
Wagner, *Global Warming Effects on Climactically-Imposed Ecological Gradients in the West*, 3, 18, 195
Wald, Study How U.S. Could Cut 28% of Greenhouse Gases, 60-61
Walsh, Comment, *A Rising Tide in Renewable Energy: The Future of Tidal In-Stream Energy Conversion (TISEC)*, 100
WARD, WATER WARS: DROUGHT, FLOOD, FOLLY, AND THE POLITICS OF THIRST, 96, 195, 198
WARD, UNDER A GREEN SKY: GLOBAL WARMING, THE MASS EXTINCTIONS OF THE PAST AND WHAT THEY CAN TELL US ABOUT OUR FUTURE, 16

Wartzman, *Can the City Save the Farm?*, 27
Washington, D.C., 121, 144, 186
Waste Management
    Batteries, 46
    Composting, 46
    Comprehensive Environmental Response, Compensation, and Liability Act of 1980 (CERCLA), 38
    Landfills, 37
    Paper cotton swabs, 47
    Plastic water bottles, 46
    Rechargeable batteries, 46
    Recycling, 37, 46
    Cherry Hill, New Jersey, 55
    Reusable household items, 47
    Shampoo and conditioner, 47-48
    Superfund, 38
    Telephone books, 51
    Trash bags, 46
Water
    Bottles, 45-46
    Conservation, 46, 197-98
    Desalination, 197-98
    Energy production, 197
    Filters, 45
    Food production, 197
    Hydroponics, 196
    Irrigation, 24, 195
    Jewelry production, 50
    Management, ch 18
    Meat production, 196
    Taxes, 197
    Use, 46-47
Waterman, Comment, *Addressing California's Uncertain Water Future by Coordinating Long-Term Land Use and Water Planning: Is a Water Element in the General Plan the Next Step?*, 199
WATERS IN PERIL, 132
WATER IN CRISIS: A GUIDE TO THE WORLD'S FRESH WATER RESOURCES, 196
WEART, THE DISCOVERY OF GLOBAL WARMING, 15
Weather, 10-11, 16, 18
    El Niño, 131-32
    Temperature rise projections 10-11, 16, 18

Weeks, *Subseabed Carbon Dioxide Sequestration as a Climate Mitigation Option for the Eastern United States: A Preliminary Assessment of Technology and Law*, 90, 198
Weinberger, *The High Cost of Free Highways*, 166
Weiss, *Drought Could Shut Down Nuclear Plants*, 97
Weisselberg, Comment, *Shaping the Energy Future in the American West: Can California Curb Greenhouse Gas Emissions from Out-of-State, Coal-Fired Power Plants Without Violating the Dormant Commerce Clause?*, 149, 188
WET GROWTH: SHOULD WATER LAW CONTROL LAND USE?, 199
WGL-Terrein, Amsterdam, NL., 121
Wheeler, *Searching for Signs of Intelligent Growth*, 142
Whitney, *Cities and Superfund: Encouraging Brownfield Redevelopment*, 35
WHYTE, THE LAST LANDSCAPE, 141
Wiener, *Designing Global Climate Regulation*, 193
Wiener, *Think Globally, Act Globally: The Limits of Local Climate Policies*, 114
Will, *Fuzzy Climate Math*, 8
Wind Energy, 92, 97-99
Wind Generators, Palm Springs, California, 128
Winters, *Carbon Dioxide: A Pollutant in the Air, But is the EPA Correct that It Is Not an "Air Pollutant?"*, 188
Wisconsin, 110
WOHLFORTH, THE WHALE AND THE SUPERCOMPUTER: ON THE NORTHERN FRONT OF CLIMATE CHANGE, 15
Wolfson & Schneider, *Understanding Climate Science*, 7
Wolk, *Cost of War Could Surpass $1 Trillion*, 167
Woltjer & Al, *Integrating Water Management and Spatial Planning*, 199-200
Wood, *Nature's Trust: A Legal, Political, and Moral Frame for Global Warming*, 15
World Bank, 87

Wortman, *No Such Thing as a Free Ride,* 183, 186
Wrye, *State Drops its Lawsuit,* 150
Yardley, *Victim of Climate Change, a Town Seeks a Lifeline,* 18
Young, *Ca. Land-Use Dispute Complicates Budget,* 149
Zaidi, *High Speed Rail Transit: Developing the Case for Alternative Transportation Schemes in the Context of Innovative and Sustainable Global Transportation Law and Policy,* 187
Zarembo, *Can You Buy a Greener Conscience?: A Budding Industry Sells 'Offsets' of Carbon Emissions, Investing in Environmental Projects. But There Are Doubts About Whether it Works,* 49
Zero Energy Houses, The Vauban, Freiburg, Germany, 123
Ziegler, Jean, 23
Zinn, *Adapting to Climate Change: Environmental Law in a Warmer World,* 195, 197